tthe~ ⌐ ⌐ ⌐ Coll

Design and Corporate Success

The Design Council

The Design Council is recognized by the Government as the UK's national authority on design. The Council's main activities are the commissioning of research projects on design-related topics, particularly stressing design effectiveness to improve competitiveness, communicating key design effectiveness messages to target audiences and developing the Education and Training Foundation, which has a broad remit covering all aspects of design education and design in education.

The Design Council is also launching a new copublishing programme with Gower in the allied fields of design management and product development. This is one of the first books in the programme. A complete list of new titles is available from Gower Publishing, telephone 01252 331551.

Design and Corporate Success

Clive Rassam

The Design Council

Gower

Published by
Gower Publishing Limited
Gower House
Croft Road
Aldershot
Hampshire GU11 3HR
England

Gower
Old Post Road
Brookfield
Vermont 05036
USA

Clive Rassam has asserted his right under the Copyright, Designs and Patents Act 1988 to be identified as the author of this work.

British Library Cataloguing in Publication Data
Rassam, Clive
    Design and Corporate Success
    I. Title
    658.5752

ISBN 0-566-07534-2

Library of Congress Cataloging-in-Publication Data
Rassam, Clive
Design and corporate success/Clive Rassam.
    p.   cm.
    Includes index.
    ISBN 0-566-07534-2
    1. Design, Industrial.   I. Title
TS171.R37   1995   94-47445
658.5'752--dc20   CIP

Designed and typeset in Great Britain by Mick Keates and Concise Artisans.
Printed in Great Britain by the University Press, Cambridge.

Contents

Acknowledgements

Many people have contributed to this book, and I would like to thank all those who gave their time and comments. In particular my thanks go to John Fisher and his colleagues at PA Consulting Group; Dr Chris Floyd at Arthur D Little; Michael Cane and Dr Chas Sims at The Technology Partnership; Dr Robin Roy, Dr Stephen Potter and David Walker at the Open University's Design Innovation Group; Professor Ray Wild, Principal, and his colleagues at Henley Management College; the staff of the Design Council, particularly Dr Colin Mynott and John Benson. My thanks also to the many designers and company executives who were so helpful, particularly Gordon Sked and Richard Elsy at Rover Group; Casey Norman at Bluebird Toys; and Sir Terence Conran, Wally Olins, Barrie Weaver, Nick Butler, John Sorrell and Bill Moggridge.

I am also extremely grateful for the help and advice of Sheila Evers at Portsmouth University and Suzie Duke, my editor.

Foreword
John Towers, Chief Executive, Rover Group

TODAY'S CUSTOMERS are faced by an avalanche of choice in every product sector, from toasters and televisions to cameras and cars. For producers, success in the market no longer rests on technical issues of quality and reliability. These are assumed to be of a high standard and indeed are now the price of entry to the shop window or showroom. The role of design in determining a product's commercial success is therefore becoming more important and is as central to a progressive manufacturer's strategy as it is in determining which product sells instead of another. In economic terms design represents the totality of elements which in combination create successful products for international trade. Good designs are successful not simply as a result of their function; they should also represent value through the emotive elements of style and character.

As this book clearly illustrates, the role of design is changing with the changing nature of business. The creative process has to be strong and self-confident enough to take its place in the new corporate cultures of multifunctional teamworking and process-driven organizations. The barriers between design, make and sell no longer apply in world-class companies which themselves are characterized by a rigorous and questioning climate in all aspects of their business. The design process is no longer the starting-point in the product cycle but a continuing process influenced by external issues and internal imperatives which, nevertheless, makes a distinctive statement about the product and the organization behind the product.

Design and Corporate Success is a practical route map rather than a theoretical guide and comprehensively pulls together the factors which impact upon design in a commercial environment.

John Towers, Chief Executive, Rover Group

1 Who needs design?

NICK BUTLER IS one of Britain's most successful industrial designers. The company which he runs, BIB Consultants, has a client list that ranges across the world. BIB is successful because through its designs it has helped to make products market leaders. These have included consumer goods bought every day and well known to everyone, such as Minolta cameras and Duracell torches, as well as industrial and medical products. When Butler is asked by people in the UK who have never met him before what he does, he tells them that he designs products. But he often finds that they do not understand what he means. Outside Britain it is different: 'When I am abroad I never have to explain what I do. If I tell someone in Japan, or China, or Korea, or Germany or Finland, they know immediately what I mean. They understand what I do. But back home people are unclear. Design doesn't seem to be part of our culture.'

Mary Lewis, founder of design consultants Lewis Moberly, which specializes in corporate and brand identity, has had similar experiences which suggest that design, although part of our culture, is not central to it. She works both in the UK and abroad, and encounters a great difference in attitudes towards design:

'The British have a nervousness about design which Continentals do not seem to share. We are not demonstrative, and design is demonstrative by its nature ... When I work abroad, it is different. People get more excited about design and our clients tend to respond more emotionally. In the United Kingdom there is a more grudging approach to design.'

These experiences are mirrored by many others both within and outside the British design profession. Professor Bernard Taylor of Henley Management College has been lecturing, consulting and writing on business strategy for many years. He has found that in Denmark (where he taught strategy to engineers for 15 years), Germany, France and Italy people are noticeably more design-conscious: 'In Denmark, you see good design everywhere you look, in their buildings, their offices, their products. The general public respect design and there is a strong craft culture. The same is true in other countries in Europe – they are interested in design. Ordinary people understand it ... Somehow we have never picked it up here that technology and design are important.'

Part of the problem is that in the UK the word 'design' means such different things to different people. Everyone has their own version of what they think design constitutes. In large part this is because the word has become dangerously over-used. People talk of designing cars, designing book covers, designing software or designing management courses. In each case the word has a different meaning. 'To design' can mean to envision, to build, to devise or to style. An even wider array of definitions exists for the word 'designer'. Like the words 'engineering' and 'engineer', 'design' and 'designer' have lost their original, narrower meaning. As the meaning of these words has become diffuse, so the understanding of what these professions really do has diminished. And what one does not understand, one does not value.

So what does 'design' really mean? In the context of products, confusion arises because the term is used to refer to the product development process or the final appearance or functional characteristics of a product, but design encompasses all these: it is everything that feeds into the development of a product and its manufacturing; it embraces technology, functional performance, appearance and management of the design and development process. As becomes clear in Chapter 4, different specialists may use the term in different ways, and often to mean primarily appearance, but that is just one facet of the whole. Professor

Design is the key to increased market share: the Ultralite Versa Notebook Computer, developed by IDEO in San Francisco with NEC's design centre in Japan, nearly doubled NEC's market share in the USA in six months, and more than doubled its share of the notebook market in the UK and Germany. (IDEO)

Stuart Pugh, in his book *Total Design*, distinguished between 'partial design' and 'total design', the latter being 'the systematic activity necessary, from the identification of a market/user need, to the selling of the successful product to satisfy that need' (Pugh, 1991); 'partial design' describes each of the many activities which contribute to successful product development. Design as a whole brings all these activities together.

This book explains the importance of design and its commercial value. It focuses on three design activities: engineering design – the research and development of new products and the processes and technology required to make them work and to manufacture them; industrial design – the development of products to meet the stylistic, ergonomic and functional needs of the market; and corporate identity design – enabling a company to present itself clearly and attractively through its corporate logo, literature, branding, advertising and interiors. Chapters 3 and 4 examine the management of product design and development, and how engineering and industrial designers work with other specialists to ensure the successful introduction of new products. The case studies which follow provide practical examples of successful product design and design management. Chapter 7 explains the influence of corporate identity design on a company's success. Whether in the manufacturing or service sector, corporate identity can be a means of binding together a company's varied products, services and activities, and reinforcing both a new culture and a new image.

Why design is important

Despite the fact that the best-known companies in the world, whose products are part of our everyday lives – Apple, Sony, Philips and many others – can attribute much of their success to

good design, many companies still fail to appreciate its power. That means failing to appreciate how influential designers can be in helping to win sales. Examples of both large and small companies which have improved their performance through the use of design are there for everyone to see, and this book includes some of the most striking, yet designers are still too often seen as quick-fix commercial artists, rather than problem-solvers who can achieve an end result that is focused, disciplined and customer-led – and which makes money. Jeffrey Bloom, a founder of Digital Audio Research (see page 85), has witnessed design's commercial value firsthand: 'Good design really does pay off. It has certainly helped us to win sales against much larger competition. There's no question that people were turned on by our novel design.' As Nick Butler also points out, 'The companies that hire designers are not philanthropists. They're doing it for sound commercial reasons.'

One of the companies featured in the case studies in Chapter 5 is the Surgical Technology Group, manufacturers of medical equipment. Managing director, Peter Gibson, is clear about the value of design: 'What we've learned is that there are clear dividends from getting the engineering design right and from putting that design into an attractive package.' With experience of a very different sector, Sir Terence Conran, whose career has been founded on the use of good design for mass-marketed products, is in no doubt about why design is essential in the 1990s: 'It really is quite dotty to see design as something that is somehow unnecessary. Everything we use is designed. The question is, do we want our products to be designed properly, that is, by the best-trained people? We wouldn't go to a doctor or employ an accountant who was half-trained. We expect them to have had a complete training. Yet in the field of product design,

Product proliferation is increasing, so differentiation is more difficult

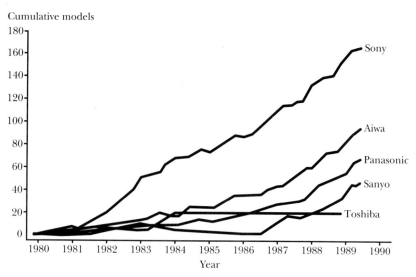

Cumulative models

Design gives products a competitive edge by differentiating them from competitors in increasingly crowded markets. (Source: Arthur D Little)

many of our manufacturing companies don't seem to apply those kinds of principles. They don't attach sufficient importance to a product being designed by a well-trained designer.' As he and many others have pointed out, design and innovation are the best way to add value to and differentiate a product, and the best way of competing successfully, when competing on price – particularly against Far Eastern imports – is not an option.

The competitive weapon

Good product design has become a major competitive weapon. Companies have moved from competing on price to competing on quality, but they must compete on design. Those who fail to understand the strategic value of design will struggle to survive.

When the basic technology available to a product is widely understood and can be developed cheaply, the competitive

option that you have left is quality and design. Particularly –
though not exclusively – in consumer markets the work of
industrial designers can make all the difference to your product.
The way a product looks – whether a manufactured or 'service'
product – is what first strikes a buyer. Politicians have long
recognized this and have used design to enforce the image of
their parties, but people in business, particularly UK
manufacturing, have been slow to take this on board.

Companies have become more inclined to use good design in
packaging and corporate identity than in product design, or in
the creation of a new service product, but this can be self-
defeating because if the product is less exciting or performs less
well than its packaging suggests, and if it does not measure up to
the image inherent in the company's visual identity, the company
will be 'found out' and the public will become disillusioned.
Using good design in only parts of the business will ultimately
lead to failure. Design brings the most rewards if it influences
every aspect of business activity.

In its battle with Gillette, Wilkinson Sword – a much smaller
company – could only compete by investing in superlative design.
Gillette's advertising spend is ten times that of Wilkinson, so
product design and styling has to be Wilkinson's distinguishing
factor. For Wilkinson's technical director, Wolfgang Althaus,
design is a strategic tool: 'Design is our main weapon in this
context.' Kenneth Grange, partner of Pentagram Design, has for
many years been an industrial designer to Wilkinson Sword. One
of the company's latest products is the Kompakt razor. The
concept was largely a design-driven initiative. The razor carries
spare blades in the handle, uses a new blade from Wilkinson's
own engineers and is deliberately presented and shaped to be
different from any of the major competitors – particularly

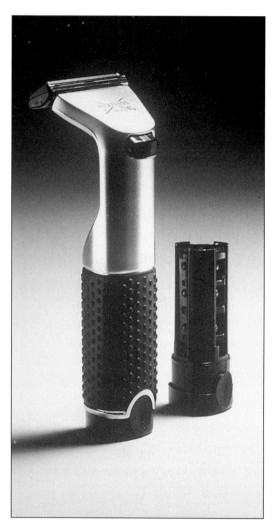

Wilkinson Sword invests in product design and styling to compete against Gillette. The Kompakt razor, designed by Kenneth Grange of Pentagram Design, is deliberately presented and shaped to stand out from competitors. (Pentagram Design Ltd)

Gillette which is macho, technical and conventional. As Grange explains, Wilkinson's product is 'ergonomic, humanistic, colourful and unconventional . . . a happy alliance of designers, technical and aesthetic.'

As is now generally accepted, one country where businesses do use design as a strategic tool is Japan. Japanese companies use design as a key part of their strategic thinking. They will often set a specific product target, based on a concept, performance features and appearance. The engineers and the industrial designers will then work towards that, and the design will become a focal point for everything else. Many designers are used to this strategic appreciation of design. Philip Gray, a director at design consultancy Weaver Associates, explains: 'Design is not there for the sake of design. It is there as a business tool to help companies sell their products better. It is not there to win awards or to put designers' names in lights.'

Design and business culture

Design imbues the whole culture of some companies. These are the organizations whose reputation and identity are intimately bound up with the excellence of their products. Dr Chris Floyd, European Director at the US management consultancy Arthur D Little, has observed this among the leading European firms: 'Those companies where products are important and where each new product has a strategic influence on all the other products that they have already, and will have in the future, tend to be the ones where design matters. An example is Braun. They sell their products on the basis that they are designed better. So if they have a bad product it would weaken their ability to sell their existing ones. So they make absolutely sure that they design their products well. Therefore design has become part of their culture. Therefore they can recruit good designers, which in turn strengthens the culture of design. So your image of yourself will determine how you see design.'

Creating a culture in which good design will flourish and be exploited is a key to successful product development and strong trading performance. When, in the mid-1980s, the National Economic Development Office commissioned James Fairhead, then at the London Business School, to investigate how the best companies in Europe and the USA kept ahead in innovation, his main conclusion was that they all used design to differentiate their products and had strong corporate cultures that were conducive to good product design (Fairhead, 1987). A report from Arthur D Little, *Management Perspectives on Innovation*, published in 1985, said much the same: 'Creating a favourable climate is the most important single factor in encouraging innovation.'

However, building an innovative corporate culture is an elusive exercise for most companies. One of the major reasons

Creating a culture in which good design will flourish, as exemplified by Microsoft, is a key to successful product development and strong trading performance. This award-winning Microsoft Computer Mouse, designed by three IDEO offices and involving extensive human factors testing, sold over two million units in the first two years of production. (IDEO)

is that many organizations are still dominated by competing groups of professionals whereas successful companies discourage interdepartmental rivalry and encourage loyalty to the company as a whole. The Rover Group is one company that has managed to build a strong corporate ethic, which has created a climate in which different professional groups can work together towards a common and coherent end (see pages 106 to 113).

Most experienced designers also emphasize the importance of the right culture. Bill Moggridge, director of IDEO, a leading international design consultancy, finds that companies that understand design have design deeply embedded in their culture: 'The managers in that company will be able to talk about design intelligently; they won't be disconnected from it.' However, he adds an important point: 'They won't just be passionate about design but also about what they're doing in other areas, for example in strategy, in marketing ...'

Managing design as a central activity

One of the messages of this book is not only that design itself is important in improving performance but that equally it is how the process of design and development is managed that is important. The effective management of design and design processes is becoming more and more crucial as companies become more international yet more sensitive to specific markets, and also as modern management tools provided by CAD and IT generally influence a wider number of functions and professional categories. The speed of change in responding to competitors and to consumer preferences is another factor. If a company's design capability is inadequate or if it is left out on a limb, then all the best financial controls or the best marketing knowledge in the world will count for nought.

The design process is now so complex and has an impact on so many staff that it ought to be central to an organization's business. James Pilditch, who founded his own design company in 1959, talks in his book, *Winning Ways*, of seeing design 'as a total activity.' He says that the world's winning companies are giving a central prominence to design not as an isolated, superior function sitting above everyone else, but in the sense that everyone else has to understand design, and therefore see how what they do impacts on the design of the company's products. Design, therefore, becomes a total activity, because at its best it involves everyone. This means that it troubles a lot of people: 'Engineers think they are designers,' because they can 'develop a product that works . . . Industrial designers think they are the ones to develop products, not engineers who have no idea what customers want.' And people in marketing, although they cannot design, have strong ideas about it too. So three interested groups of people are all likely to think they know what makes a good design, and, what is more, will have very different ideas of what a design really involves.

Trying to bridge these differences and get different professional people to work together in teams so that their unique competences can be used properly is fairly difficult for most companies. Those companies who have succeeded admit that it can be a lengthy process.

Management attitudes

One of the greatest problems influencing the use of design is that it has been managed badly or rather it has often not been managed at all. On the one hand, Britain has some of the best industrial design and engineering talent in the world, but on the other hand, particularly in the consumer products industries,

British companies tend not to use these skills well. Some blame is always attributed to the fact that engineers are not good at working with those in Marketing or that design simply costs too much, but there is much more to it than that. The problem starts at the top, as it almost always does. Many senior managers and directors have insufficient appreciation of what design means even when they employ designers – be they engineers, electronics experts, ergonomists or stylists – in their companies.

The view of David Carter, Professor of Industrial Design Engineering at the Royal College of Art and Imperial College, and chairman of the DCA design consultancy, is that many senior managers do not understand the process of design and its complexity. They therefore often fail to understand its financial value. In particular, he says: 'A lot of people don't understand that something can be developed from a blank piece of paper, and that what you put on that paper can make all the difference to a product and therefore your profits.'

Design is about taking risks, making mistakes and moving forwards. It is also about working unofficially and informally with others in the organization – in other words it is much more haphazard than senior managers realize. Whereas project managers tend to understand this, senior managers and directors whose lives have been shaped by other disciplines and other experiences often will not. However, design can nevertheless be efficiently managed. Successful companies have found that designers and engineers need the freedom and flexibility to experiment and make mistakes, but that this must be part of a well-understood and well-managed development process. The experimental and iterative nature of design means that an environment in which ideas can be expressed freely must be fostered, but not at the expense of structured management.

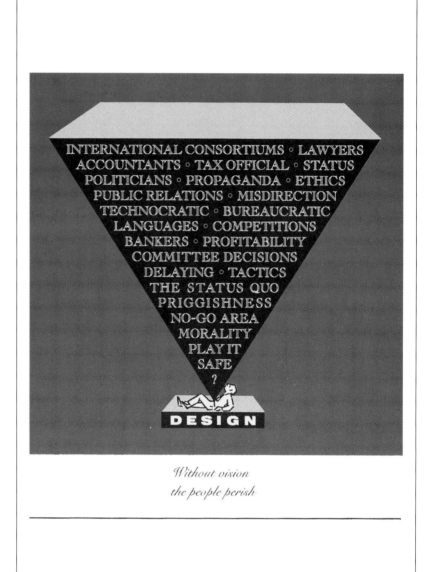

The status of design? This illustration was designed by James Gardner for the July 1994 issue of the *RSA Journal*. The front cover for each monthly issue is designed by a Royal Designer for Industry. James Gardner was elected an RDI in 1947. (RSA)

The status of engineers and designers

Another problem is status. In the UK both engineers and designers lack the kind of status that most other professions enjoy in an organization. Even if they are represented at board level, their status will be less than that of the finance or sales director, yet the latter are likely to have little firsthand experience of project management. Engineers' and designers' lack of status can also undermine the multi-skilled team philosophy which depends on the equality of team members for success. If one team member, for example, the financial representative, has a higher status than the engineer or the designer, the value of teamworking is lost (see page 52).

The lower status of a designer compared to a marketing person is often apparent and can have a significant effect on how design is valued by a company. Dr Stephen Potter of the Design Innovation Group at the Open University makes the point that often the success of a product is laid at the door of the marketing people when it has been the designers who have been responsible. As an example he cites the mineral water Highland Spring. More advertising alone failed to raise its sales. However, when the bottle's labelling was redesigned and then followed up by more advertising, sales increased. But, he maintains, it was the marketing people who won the plaudits not the designers. In a proper team-based culture, the whole team rather than one department would be credited with success.

Short-termism

At one level – as customers – we are all conscious of design in the sense that it clearly determines what we buy. But at another level – as business planners and strategists – many of us hardly allow design to enter our thoughts. Professor Bernard Taylor of Henley

Management College finds that generally in the boards of UK companies design is rarely mentioned. What he hears about most is related to operational efficiency, sales and financial questions: 'They're worried about costs and overheads and cash flow. Some of them are concerned about survival. They don't talk much about new products, yet – they haven't the resources. So design doesn't exist for them.'

Every experienced designer is aware of the financial attitudes that obstruct the acceptance and understanding of design in boardrooms of UK companies. As Kenneth Grange explains: 'If you look at most UK companies they are driven to make money now. They are run by financial and marketing people, which cramps the expansion of a lot of design, because both kinds of people think about the next twelve months not the next five or ten years. I've had a succession of UK clients that have been taken over and that have had their research and design departments axed immediately, purely for short-term gain. I don't see that with my overseas clients.'

Certainly there is a lack of risk-taking in many companies, particularly publicly quoted ones, and often decisions are governed by financial short-termism. Companies have to look for quick returns, and many designers find that clients will give all sorts of reasons other than the real ones for not going ahead with a new design or a new product. This is the experience of Michael Cane, at the technology consultancy The Technology Partnership: 'We have recently shown the same piece of technology that we have created to two companies, one in the UK and then one in Germany. For the UK one, the technology could revolutionize where they are in the market, but they said that they couldn't develop it because they couldn't see how they would sell it. When we took it to a German company, they said they were interested, that they would look at the market, develop our concept to a workable prototype and test market opinion. Then they would make a final decision, which would mean making our product under licence.' Cane believes that the difference in response is partly based on a failure to understand the process of design and, for example, how a prototype is assessed against market research, but also on unwillingness to invest money in a new technology.

Many companies still think in terms of the lowest common denominator – the least that they can do to stay in the market. Many are still entirely cost-driven and are very wary of investing in new design. Decisions are not governed by technology, or even market demands, but by short-term financial objectives. Barrie Weaver of industrial design consultancy Weaver Associates points out the fundamental contradiction which holds companies back: 'Design looks forward. If you don't have people thinking about the future and what the market needs you won't have a company in the long run at all.'

Companies need design for survival. John Sorrell, co-founder of design and identity consultancy Newell & Sorrell, and chairman of the Design Council, feels that the City needs to be shown how important a role design plays in companies' success: 'Design is one of the components that is going to make us competitive in the next decade.' The Design Business Association is already demonstrating the value of design through their Design Effectiveness Awards, which aim to promote understanding of the commercial benefits of investing in design, and demonstrate that the effects of design can be analysed and measured. The next chapter sets out financial arguments for investing in design and shows how design and effective management of the design process can strengthen performance and improve profitability.

2 Design and financial success

IN THE POPULAR BBC television series, *Troubleshooter*, Sir John Harvey-Jones deplored the small sums companies invest in design. Elsewhere he has commented: 'Good design enables a company, however large or small, to compete profitably in world markets with quality products at competitive costs.' (Constable, 1991) The financial benefits of good design and good design management are still questioned by many companies; they doubt whether design pays or, even if they are aware of the sales that better design can stimulate and the cost-saving potential of better process management, they worry that the investment needed may be too high or that they will have problems introducing best practice techniques. Yet many successful British business people can recount an experience that has convinced them of the power of design and the financial rewards it brings, and in countries such as Japan, Germany and Italy, design is seen as an intrinsic element in product and corporate success. Of course, design is not the be-all and end-all of success, for there are other factors that affect a company's performance, but without excellence in design and design management a company makes itself extremely vulnerable.

Giving design a chance

The financial benefits of design are proportionate to its position and influence in a company. If design is used as a strategic tool, as it usually is in the most successful companies, and if, in terms of status, it is strategically positioned within an organization, then it can provide considerable financial benefits. Design will pay off handsomely if it is used in the right way and in the right context. In many ways design is like a plant seed: its lasting value depends on when it is planted, in what soil, among what other plants and then how it is nurtured. Success stories of small- to medium-sized firms researched by the Design Council show that a key factor has been attention to strong product or service design, and in every case there has been a deep commitment to getting product details right for market needs (Farish, 1995).

For many Japanese companies, design is where business plans begin. They invest large sums in their product design and scour the world for the best industrial designers. Japanese companies not only use design as a strategic competitive tool, they also position design strategically within their organizations. There is a good reason for this, explains Japanese designer Nobuoki Ohtani: 'At the end of the day, to make a design concept happen is not in the designers' hands. It is a management decision. So the designers have to be at the heart of the decision-making process.'

Many firms go wrong because they do not make design a strategic activity. It is peripheral to them. Most manufacturing companies are either sales led or financially led, yet the source of new products is design and development (even if the function is not carried out by designers) and the source of many of management's problems comes from mishandling that design function. In other words, how design is managed has a financial impact.

Design and export performance

Most of the winners of design awards are also high exporters. This supports the view of Dr Chris Floyd, European Director at Arthur D Little management consultants, that 'those who are good at design are usually good at everything else.' So a company that has a high all-round competence is obviously likely to do well as an exporter.

The relationship between design success and export performance has been noticed by a number of academic researchers and writers. Researchers from the Design Innovation Group (DIG) at the Open University and the University of Manchester Institute of Science and Technology (UMIST) have undertaken a series of studies on design and business success. Since the mid-1980s, they have looked at over 100 companies based in Europe, Japan and Canada. The aim, says the DIG's coordinator, Dr Robin Roy, has been 'to isolate management policies and design practices pursued by commercially successful firms producing "well-designed" products.' The industries they have examined include plastic products, office furniture, electronic business equipment, bicycles, motor vehicles and domestic heating equipment.

Roy explains that the 'design-conscious' firms were defined as those that had won design awards or citations from reputable design bodies and which were considered by competing firms to produce well-designed products. Not only did the researchers find a correlation between design awards and profit margins, but they also found that those firms which had over 20 citations from the Design Council exported a significantly higher percentage of their sales than other, less design-conscious companies they examined. Export success was not a criterion for selection of the award-winning companies, so the link between design awards and export success was not necessarily a foregone conclusion.

What the DIG researchers also found was that the best-performing companies (in terms of, for example, profits and sales growth) took a comprehensive view of design, both in its concept and in its role within their organizations. Some firms adopted a rather narrow definition of design, defining it principally either in visual or technical terms. Others – and these proved to be the more successful – saw design from many angles, such as fitness for purpose, production efficiency, profit and styling. In his chapter 'Product Design and Company Performance' in the book *Design Management: a Handbook of Issues and Methods*, where he comments on these differing attitudes towards a definition of design, Robin Roy says: 'Having a broad understanding of the meaning of design may not seem very important, but the firms that did have such an understanding also realized how design decisions did not just determine concept, form and performance, but also influenced all the other factors that contributed to a product's competitiveness in the market.' (Oakley, 1990)

A similar but more detailed view has come from Dr Ughanwa and Professor Michael Baker in their book, *The Role of Design in International Competitiveness*. One of their approaches was to question a large sample – more than 90 – of Queen's Award-winners in manufactured exports, asking them about the importance of design, research, marketing and strategic factors

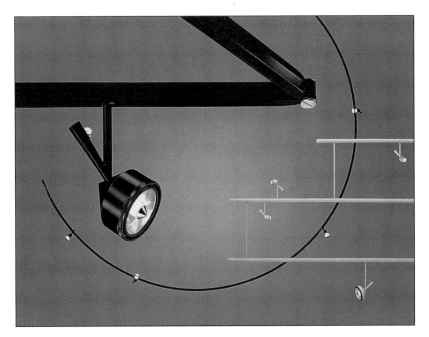

Firms recognized as 'design-conscious' tend to be higher exporters. Concord Lighting's Infinite Lighting System, a British Design Award-winner, is based on an idea from product designer Terence Woodgate. It is sold in the UK, Europe, the Far and Middle East and Australia. By its nature it can expand and evolve to meet future market needs. (Concord Lighting)

in their success. Anyone familiar with good design practice in its widest sense would see that most of these high-exporting firms were in fact doing all the right things – or at least their responses suggest that they were. These firms were using design for competitive export advantage (see also page 52).

Design and company costs

It is widely recognized that the design of a product dictates its costs, but the implications are still frequently ignored. Although the early design and development costs are low compared to the costs of the whole product development cycle, the impact of this early work has deceptively far-reaching financial effects and it is dangerously easy at that early stage to be lulled into a false sense of optimism. Dr Steve Bone, who specializes in technology management at PA Consulting Group points out: 'The concept stage is crucial because it is then that the basic design is established. In laying down the concept you are almost picking the production processes, the materials, the suppliers – it's virtually all decided at that early stage, and it is very difficult to change it from then on.' Dr Colin Mynott, former Director, Industry, at the Design Council explains further: 'By the end of the design and development process over 80 per cent of the costs may have been pre-ordained. Beyond that stage, production can typically create only 15 per cent of additional value.'

This explains why so many projects that seem to be feasible at the conceptual stage turn out to be uneconomic: the full development and manufacturing costs and the consequences of the chosen concept are not accurately calculated. What happens often is that a project is given the go-ahead, but the post-design

Design dictates costs

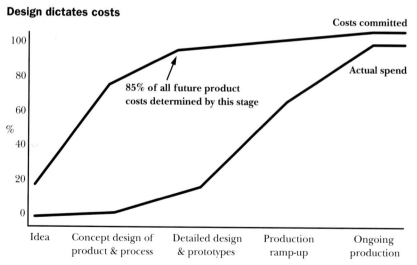

The design of a product dictates its costs: the impact of early work has far-reaching financial effects.

Design definition and reduction of technical and commercial risks

Approximate % of total project expenditure	Definition established (milestone)	Parallel tasks
Start	Marketing requirement	Market research
5-8	Feasibility demonstrated	Customer research
10	Design brief agreed	Manufacturing, Quality Assurance
20	Concept design agreed	Customer research, costings
30	Detailed design agreed	Update of plan, tooling estimate
55	Prototype evaluated	Marketing, costings
65	Design for manufacture	Tooling, manuals
80	Pre-production evaluated	Field trials, packaging
100	Product launch	Production

Increased design definition should ensure that, as project expenditure increases, technical and commercial risks are progressively reduced. (Source: The Enterprise Support Group)

costs – most of which are dictated by the design – are underestimated. This has happened time and again with major UK engineering and electronics projects. The fault does not usually lie with the engineers and designers but rather with senior management, the structure of the organization and the effect that it has on the management of the whole product design process and associated risks.

A study by Berliner and Brimson in 1988 showed that investment in the design concept has a far higher return than investment in new manufacturing methods or manufacturing strategy. This may seem hard to believe, but, given that so many companies are manufacturing less and less of their output, it is not that surprising. Senior executives at Ford have confirmed this conclusion by saying that 'design decisions are ten times as effective as production planning decisions and one hundred times as effective as production changes in reducing costs and improving quality.' (Roy et al, 1990)

Yet this leaves the question of how those costs are going to be analysed. Mynott argues that the traditional accounting view of a company's manufacturing costs makes a simplistic analysis and therefore generates the wrong answers. In the traditional view there are controllable and non-controllable costs. Among the controllable costs are 'certain overheads and direct costs, which in practice relate to employment costs.' He maintains that conventionally the controllable costs are considered to be the area where improvement in efficiency can most readily be effected to improve company profitability. However, in reality the so-called 'non-controllable costs', which include raw materials and bought-in components, 'are a far better source of achieving better cost-effectiveness.' The 'non-controllable costs' are 'largely controlled by the way the product is configured at the design

Analysis of manufacturing costs

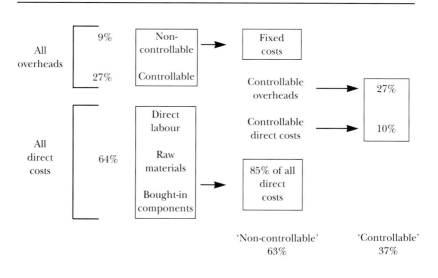

Traditionally 'non-controllable' costs, such as raw materials and bought-in components, can be reduced by the way a product is designed, and less complex designs can reduce plant costs.

stage.' He adds: 'The cost of production equipment and power consumption, the size of the plant needed and the complexity of the manufacturing systems are all a direct result of the way the product is designed. These costs are under the control of the design department.'

Profit versus time

The timeliness of design also has financial implications. If a product is launched late a significant proportion of potential sales are lost. This has become ever more important as product lifecycles have shortened. Most of the companies featured in this book made or were making products for quite specific launch times, either because the product was selling into a seasonal market (such as garden equipment and toys), or because it was to be unveiled at a major exhibition (audio electronics and medical equipment) or because the product was timed to surprise the market ahead of possible counter-moves by competitors (medical equipment and micro-light aircraft). In the jargon of the management gurus these companies had a window of opportunity, which if missed could never be regained.

In the cases of DAR and the Surgical Technology Group (pages 85 to 95), award-winning design was coupled with a fine appreciation of the customer's needs plus a timely launch. Both products were developed very quickly – in less than 18 months. The power of the design was strengthened by management capability to deliver a product that was honed to a specific market for a specific time. The experience of these two companies supports what every piece of academic research has discovered: that competence in good design has to be aided by the high performance of other management functions.

One of the cardinal laws of project management is Murphy's

Time to market affects both product life and market share

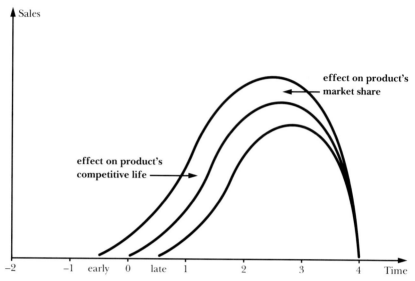

Companies must understand the financial value of time in product development projects. Launching late will often be more damaging financially than launching on time with increased development costs. (Source: Arthur D Little/Cambridge Consultants)

Law: projects almost always cost more than expected and take longer than anticipated. There comes a time in most development projects when costs are rising, the accountants are getting tetchy and the sales people are screaming as the launch date grows nearer. The solution is usually a compromise – trying to balance the immediate financial and sales needs of the company. Colin Mynott cites the example of two companies faced with sharply rising developing costs. In one case the company spread its increased costs over a period of months for internal accounting reasons and brought out the new product six months late; the other company allowed its development costs to spiral upwards by 50 per cent over a short timescale enabling it to launch just about on time. The first company lost 33 per cent of its lifetime profits on the new product; the second company reduced lifetime profits on its product by only 3 per cent. Launching a product late can often have a far worse effect on profits than, within limits, increased development costs.

Cost-saving management

One of the areas in which the management of design and development has a major financial impact is 'design for manufacture.' This aspect of design is now taken much more seriously by leading companies, but it requires that design is put centre stage or at least put on an equal footing with other management functions. Creating multidisciplinary teams is one of the favoured routes by which the manufacturing costs can be brought to bear on the design process (see Chapter 3). The consideration of manufacturing early in the process can dramatically reduce costs, for example through reducing the number of late changes when the design is released to manufacture.

Among the many ways in which companies can reduce design costs – through streamlining machining, assembly and use of materials – two stand out: designing a product around a series of modules and using a higher percentage of standard, bought-in components. Modular-based design has two advantages: it enables designers to design several variations on one theme to suit an array of different customers and it facilitates incremental changes to be made for successor products. Modular design assumes a flexible manufacturing system so that costs are kept to a minimum all round.

Modular-based designs have become popular among Japanese car-makers, while in Britain British Rail Engineering (BREL) designed a modular-based carriage in the 1980s that allowed the manufacture of carriages with different dimensions and interior fittings all on the same production line. The BREL design was also exported to a number of countries around the world.

Successful companies have streamlined their manufacturing in this way and often actually manufacture very little themselves. They therefore have fewer 'non-controllable' costs than previously. The practice of buying in more supplies rather than making them in-house is growing among leading Western manufacturers, so the designers have a great opportunity for keeping the price of supplies down by designing products that will use the most cost-effective components yet remain functionally and ergonomically appropriate.

The potential for cost savings through the use of modular designs or the skilful selection of component suppliers is considerable, but this does require designers to have a close relationship with other departments in a company. For example, to create a worthwhile modular design it helps if the designers have some notion of how the marketing people see the further development of a particular product, otherwise the set of modules will be inappropriate or at best under-utilized in the long term. So, once again, the expert use of design at an early stage can have substantial benefits but only if the design process is fully integrated into the company structure.

The value of non-price factors

In any discussion of international business the conversation sooner or later turns to the importance of non-price factors and value for money rather than cheapness in securing market advantage. These factors include uniqueness, reliability, ease of use, quality and after-sales service. The significance of non-price factors in helping companies to gain market share is now widely recognized. The implications for design management are very clear because these factors are largely determined by design –

principally the technical performance and the appearance of a product, which includes the extent to which visual and external features improve the way it functions.

Good market-driven design is a key element of success in sectors ranging from electronics, computing and toy-making to wallpapers and engineering. Successful companies are not

Good design: key success factors

1	Performance:	does it do the job reliably?
2	Ergonomics:	is it easy and comfortable to use?
3	Aesthetics:	does it look good?
4	Maintenance:	is it easy to maintain and service?
5	Safety:	is it safe?
6	Manufacture:	is it easy to make?
7	Price:	is it good value for money?

These factors are largely determined by design and are crucial to securing market advantage. (Source: Priestman Associates)

necessarily original or highly innovative, but what often marks them out from others is an exacting attention to detail in the design of their products. They may be largely imitators, as the Japanese used to be, but they have striven for excellence in design to outgun their competitors. For these entrepreneurs design is a strategic weapon, without which they know they would not survive.

Research by two American academics, Kravis and Lipsey, published in *Price Competitiveness and World Trade*, showed that in US/West German trade the price advantage accounted for only 28 per cent of trade success, whereas 47 per cent was derived from product superiority or uniqueness. So the non-price factors were dominant. In the UK, research by Schott and Pick published in 1983 came up with a similar conclusion. They showed that in the UK's home market non-price factors were responsible for up to 80 per cent of British imports. In other words UK consumers were not making their purchases solely on the grounds of price.

When the researchers examined UK export performance they found that non-price factors only accounted for 45 per cent of British exports. This study confirmed a report by David Connell for the National Economic Development Office in 1979, which argued that UK companies were trying to compete largely on price whereas foreign companies put as much if not more emphasis on non-price factors. One result of this approach has been that for most of the 1970s and 1980s the unit value of UK manufactured exports has tended to be lower than the unit value of UK manufactured imports.

The importance of design-influenced non-price factors is especially obvious in the consumer sector. In the major consumer sectors, such as motor cars, cookers, washing machines, hi-fi systems, televisions, furniture, sportswear, electric heaters, small electrical appliances and toys, the decision to buy is strongly influenced by the product's design, especially performance, ease of handling and styling. In the UK home market all these sectors have been heavily penetrated by imports. More than 70 per cent of electronic consumer goods, toys, sports goods and motor vehicles are now imported. Imports have sometimes gained a foothold through price rather than through technical or ergonomic qualities, but the weakness of UK design, especially stylistic, industrial design, has been an important factor.

Medical equipment is a sector in which design has become paramount, firstly because medical technology is increasingly complex and good design is needed to bring all the various technical elements together harmoniously; and secondly because this sector has been targeted by Asian countries who are able to produce reasonable products at very low prices, which means that Western medical companies have continually to upgrade their designs.

The importance of concentrating on non-price factors is underscored by Sir Terence Conran: 'If you're basing your selling strategy mostly on price in an international market, you won't stay in business for very long because there will always be someone who will sell for less than you do.'

The benefits of design investment

Many companies feel that the pay-off from investing in design expertise is low. They say that they cannot afford it: the outlay would be too high and the rewards would be unquantifiable. This attitude is especially true among small- and medium-sized companies (SMEs). However, what research there has been into the effects of design investment in SMEs has shown that there is a better financial pay-off than is commonly believed.

Between 1990 and 1991 a team of six researchers from the Design Innovation Group researched the impact of government-subsidized design investment on 221 firms (Open University/ UMIST Design Innovation Group, 1990). These firms had received a subsidy – between 1982 and 1987 – to employ a design

The benefits of design investment

Type of design expertise	Mean total project cost*	Mean payback period from launch	Percentage profitable
Product design	£54,100	15.9 months	97%
Engineering, engineering design, industrial design	£64,000	15 months	86%
Graphics/packaging	£37,900	11.5 months	100%

*Including all research, design, development, marketing etc.

There is a better financial pay-off from design investment than is commonly believed. (Source: OU/UMIST Design Innovation Group)

consultant. The companies concerned were a cross-section of UK manufacturing industry and most of them had had little previous experience of using design consultants. Given the inexperience of these companies in the use and management of design expertise the results of the research were encouraging and much better than expected.

The researchers found that about two thirds of these funded projects were fully implemented – they resulted in a marketable new product. Of these implemented projects, '89 per cent paid back their total project investment and made a profit. Nearly 90 per cent of these successful projects achieved the payback period of three years generally required by small- to medium-sized companies; indeed the average payback period was 15 months from the launch of the product.'

The researchers added that 'where it was possible to calculate the increase in sales resulting from the redesign or repackaging of an existing product, this averaged 41 per cent. A quarter of projects enabled a company to enter a new home market while 13 per cent resulted in new or increased exports. A further 36 per cent had other international trade benefits, largely through strengthening British goods in the UK market against competition from imports.'

The average investment in these design projects had been £60,000, which suggests that these firms had reaped considerable benefits for a relatively small outlay. The main difference that the investment had made to the design of companies' products had been to improve their specification and performance, and to enhance their styling and visual appearance. Although non-design factors had sometimes played a part in the improved performance of these companies, the researchers concluded – after lengthy interviews with company managers – that in most cases better design and development work had been the major influence. Stephen Potter, one of the researchers, reflects: 'This study shows that good design undoubtedly does pay and that investment in design is a much lower-risk activity than most managers think.'

Potter and his co-researchers also looked at the factors that inhibit good design and therefore limit its financial benefits. Dr Roy points to two factors: poor market research and an inadequate brief. He argues quite rightly that market research 'is not terribly easy to do,' and that often the research is hasty because the development project itself is being rushed through. To do your market research well and to define the new product brief properly requires time, so if the resources are not available 'everything is done skimpily.'

The OU/UMIST group have also looked at the financial effect of good design and good design practices in firms that have been recognizably successful in their sectors for some time. Their earlier research on 'design-conscious' firms indicated that design does show up positively in a company's balance sheet. When the researchers focused on a sample of UK heating, furniture and electronics companies, they discovered that the 'design-leading' companies – those who had a high number of Design Council citations and awards and a high evaluation by competitors – had 'significantly higher profit margins and return on capital than the remaining UK firms in the sample.'

Companies that understand design and manage the design process well are usually aware of its financial benefits. The research carried out by Ughanwa and Baker certainly demonstrates this. When they asked their sample of Queen's Award-winners in manufactured exports to explain why design was important for them, the answers show that they were definitely conscious of the financial rewards of having good design practices. The authors say that: 'Of the four major factors selected by respondents as best encapsulating the concept of design, "making products that make profit" was ranked as most important by 51 per cent of the sample,' and 35 per cent of respondents said that 'making products that sell' was most important. These firms are seeing design as a financial tool as much as a sales tool.

Focusing on the long term

The problem for most UK companies is that they are working within very tight financial restraints and timescales. The City institutions that invest in publicly quoted companies have little understanding of the long-term role of research, design and development in securing a company's place in an international market three years hence, and the banks which provide capital to the unquoted companies are staffed by managers who have never run anything except a bank so they tend not to understand the power of design either. Moreover, while design costs may be small, investment in other areas of management such as research and development are likely to be high; so getting a return on investment in good design is often dependent upon other types of investment which may not be forthcoming.

The view of many consultant designers is that financing design for UK companies is a problem, and some have experienced that this is particularly true of some of the larger, more acquisitive companies.

Taking the wider view

The financial pressures upon UK companies may explain why they tend to take a narrower view of design than their foreign counterparts. When the OU/UMIST researchers looked at their sample of furniture, heating and electronics companies, they found that 'two thirds of firms stated their objectives in terms of profit, sales growth or market share targets,' but not in terms of quality or design. 'Only about one in six firms expressed their main objective in terms of product excellence or good design.' The researchers also found that foreign firms were more likely to mention 'value for money' as giving their products a competitive edge than the UK firms. When the researchers examined a larger sample of companies across even more industries 'it was notable that the foreign firms tended to attribute product competitiveness to more factors than their UK rivals. In particular, technical performance was crucial to all the foreign firms, with a high proportion also including product quality, delivery and after-sales service.'

Design will have a financial impact on a company if a philosophy of design pervades that company, if the design function is allowed to contribute to other functions and if design really adds value to the company's products. Some of the influence of design may not be immediately quantifiable, but that should not be a reason for squeezing the design budget. Good design earns its own price and the high-value product can be the high-profit product.

3 Managing product design

▶ *Getting the concept right* ▶ *Evaluating risk* ▶ *Managing information* ▶ *The role of marketing* ▶ *Analysing customer needs* ▶ *Project teams* ▶ *Company culture* ▶ *Involving suppliers*

THE MANAGEMENT OF product design is often a black hole. In many companies, there are few people who understand the process of design; the overall responsibility for it is diffused and when a new product does manage to emerge out of this unprepossessing climate no-one quite understands how it has all happened.

It need not be like this. Properly managed design is an integrated process and an integrating one: successful design and design management depend upon the contribution of different professionals in a company and the process of design can therefore serve as a coordinating focus of a company's activities. In many of the best, most enterprising firms the design function has in effect become the hub of the whole organization. That way design gives a company coherence and also provides a corporate meaning for employees. One saw this in the 1980s in Silicon Valley, where scores of start-up companies were so obviously built around a product design, and this focal point generated tremendous enthusiasm among the employees.

If companies are to manage design properly it needs to be at the heart of their operations and needs to be reviewed regularly. One of the problems with the muddling-through approach to design is that no lessons are learned, because there is little information from which to draw conclusions. Hence there is a lack of understanding of what has happened in a product development project, and a lost opportunity to do it better next time: the process must be mapped, modelled and reviewed.

Managing the function of design is not easy. There are several reasons for this. Firstly, design cuts across many other functions and professional territories; secondly, it is deeply political and therefore its practice is highly influenced by the culture of an organization; and thirdly, the activity of design is affected by almost every management technique that managers adopt, whether it is Total Quality Management, or Concurrent Engineering or some new financial tool. This last point is rarely understood and that is because most branches of management are too often seen as discrete activities, compartmentalized one from the other. A fourth reason why the design process is difficult to manage is that it is increasingly being split between people inside an organization and those outside it, be they suppliers, commercial partners or consultant designers.

Getting the concept right

What is the most difficult part of the entire design process to get right? The overwhelming view of senior managers, management consultants and business academics is that the messy part of the design process – 'the fuzzy front end' as Smith and Reinertsen call it in their book, *Developing Products in Half the Time,* is not so much transforming an idea into a manufacturable product or matching marketing needs with design ideals but actually creating the right concept in the first place. Howard Biddle, deputy managing director at Cambridge Consultants explains: 'The thing that will determine a product's success is how well it

Getting the concept right

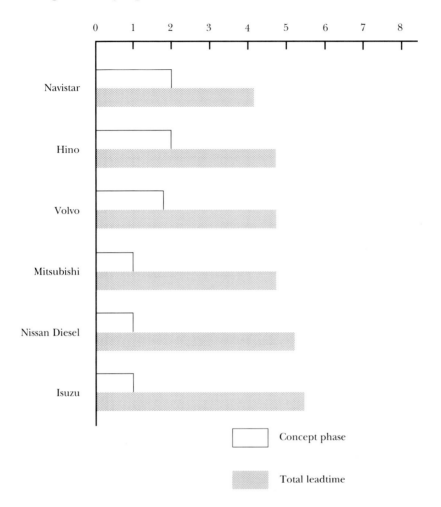

Companies that devote more time to planning and conception are able to shorten overall leadtimes. (Source: Arthur D Little/Cambridge Consultants)

has been specified at the beginning.' Part of the secret of getting that concept right is having as many possible options and as much information as you can: 'This means understanding all the technologies available, brainstorming ideas and then feeding in customer needs and the various manufacturing and marketing constraints into the ideas that have been brainstormed.'

Another important factor at this early stage is understanding how a company's own constraints and capabilities will influence its product design and development. PA product design specialist Michael Paton has found that companies often fail to think through what their customers really want and what response is best suited to their expertise: 'They don't think about what their core competences are and how they can integrate those competences with their product design. If they can do that then they can work through questions such as design for manufacture, design for the market and design in terms of visual positioning.'

Design can be seen as a bridge between technological expertise and customer needs – what some Japanese manufacturers call 'seeds and needs' – both influenced by productivity and marketability factors. When Toyota initiated the Lexus range of luxury cars, the product strategy was built around use of existing manufacturing lines because their manufacturing processes were a core competence. This demonstrates why design has to be a central activity and why it has to be closely connected with other functions.

Planning the development of a new product is also difficult to get right. PA have found that once the concept is agreed, companies find that the hard part is developing that concept – putting together a total product definition. This foundation work is often underestimated and underdone, which results in the product definition still being rationalized and thought through when the actual physical development has started. Too much change takes place and the goal posts start shifting, which means toing and froing with Marketing, causing increasing frustration and continually extending timescales.

Evaluating risk

Another part of this early process is examining the risk associated with developing a new product. John Fisher, technical director at the PA Consulting Group, believes that risks must be faced head on and balanced against the commercial opportunities. A formal risk evaluation procedure which is invoked during the project is essential, because there is a risk that the technology will not work, that a competitor will move in or that the end customer expectations will change. In his experience there is a tendency in some companies not to talk about risk because it implies uncertainty and therefore makes people uncomfortable.

Managing risk requires an evaluation of the ultimate goal and the time horizon for that goal. The focus should not be on whether something is a 'good design' in any absolute way, but on the positive or possible negative impacts on the profitability of the business in the long term.

Setting a long-term goal can work wonders for clarifying design objectives. In this respect one key mistake that companies make is that they will have a concept and try to achieve it 100 per cent in one go. Very often it is possible for a company to

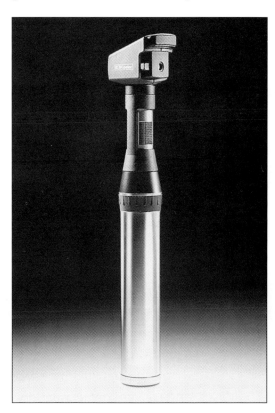

The Keeler Vista Retinoscope was designed and developed by PA to allow efficient, cost-effective manufacturing, to be easy to use and to reflect high quality visually. (PA Consulting Group)

launch a compromise product – one that is only half way to what it is aiming for – and then to bring out the perfect model later. This can greatly improve profits in the long term. However, it is impossible to generalize and PA has encountered dilemmas with different financial and market conditions, so the right company response in those cases was to unveil the new product later than expected but further advanced technically. What both cases have in common is a sensible evaluation and management of risk.

Managing information

What differentiates the successful company from others is the way that it synthesizes information from within and outside the organization. How many reliable sources of new information does it have? Does it let vested interests get in the way of receiving and interpreting that information? Is that information fed to the right people at the right time? The quality of information from the marketing and manufacturing departments clearly affects the quality of a company's product designs. As Japanese designer Nobuoki Ohtani points out: 'To stimulate designers' minds, you should give them as much information as possible, saturate them with information, and then you can have a very creative solution.' Because engineers and designers have often been sidelined both organizationally and geographically, the flow of information between Design and Manufacturing – and Marketing – has often been poor. It is the responsibility of senior management to remedy this.

Increasingly, information-gathering is a key factor in setting strategy and managing design. This point is emphasized by Alan South, group leader in product design at Cambridge Consultants: 'Good design comes out of synthesizing the right product from information on your competition, markets and technologies. It is this process of synthesis and selection that companies so often get wrong ... Sometimes decisions that will have a major effect on the market success of a product are made by poorly trained people working with inadequate information. It is depressing to see avoidable market failures and to think of the lost opportunity and the waste of all that engineering and design effort.'

The tendency for companies to make major changes to their design half way through development, often because they have not 'synthesized' the relevant information in the first place, is widespread. Obviously some changes to the initial design are inevitable; what is at issue here is the scale of alterations and the impact that they have. The more changes that are made, the greater the pressures on the manufacturing, the suppliers and the launch date.

Late changes and black holes

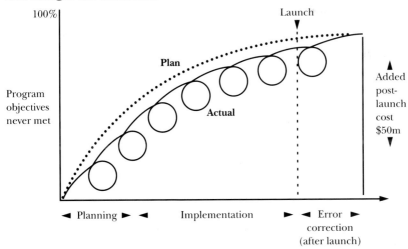

'Such a painful new model launch – never again!' – CEO of a European truck manufacturer. (Source: Arthur D Little Ltd)

The accent on thorough groundwork is very much a Japanese and to some extent a German approach. It is now generally accepted that in Japan companies spend about 60 per cent of their time thinking about the product and the rest making it, whereas in Europe the tendency is to spend about 40 per cent of the time designing the product and the rest of the time trying to get it right.

The role of marketing

It is a commonplace remark that good design begins with the market. But who is interpreting that market? Where do you draw your market information from? Does market research have much value? What happens when market research or just 'market feel' clashes with designers' knowledge and instinct? These are some of the controversial aspects of the marketing function and its relationship with design.

The strength and competence of a company's marketing department impinges directly upon the efficacy of a company's design capability. If the information emanating from the marketing department is flawed then the next new product design is likely to be deficient as well. Terence Conran holds a strong view on this: 'Many companies' market research is useless. There is too much of it and it is too statistical. What we (designers) do is to talk to potential users of a new product, and then we'll interpret our findings into something that is tangible, probably by making a model.'

In his book, *The Design Dimension*, Christopher Lorenz devotes a whole chapter to what he calls 'The Marketing Conundrum'. Drawing on the work of the US business guru Theodore Levitt, he pulls apart the work practices of most marketing people. In particular he draws attention to the limits of market research and also to the statistical bias of the marketing profession. Marketing people, he says, 'need to be strategic thinkers and good communicators,' but 'their training and work experience tends to turn them into over-numerate technocrats.' He adds: 'By no means all of them receive a formal training; for those who do, it is techniques which are stressed, not human behaviour, motivation and communication.' In other words the marketing people are quite likely to misunderstand or even miss altogether important signals coming from the market.

There is often a difference between what the marketing people say is needed and what the market really needs, sometimes because marketing people are too cautious. Lorenz's main thesis (as it relates to marketing) is that the marketing function only succeeds if it is truly a 'window on the world'. In other words, if it observes all sorts of social and environmental information bubbling away in the world outside the corporation. That is the context in which marketing ought to influence the genesis of a new product, but, as Lorenz points out, this approach is much more widespread among Japanese than Western companies.

One way to have a window on the world is to have a direct relationship with customers. There are two points at which marketing information traditionally impacts on product design: at the initial concept stage and later on before the product is put into production. Where marketing can have a major role to play is in this second phase – in the market-testing of a new product. Almost all the small firms featured as case studies in this book test-marketed their latest products on a carefully selected number of potential users. Having trusted testers is obviously a problem, because if word gets out that a company is developing a particular product, at best the element of surprise is lost by launch day, and at worst a competitor can come in with a spoiling product alternative.

Much more difficult is to test a product when the technology intrinsic to that product is still being developed. PA's approach is to create a product that looks like the one contemplated but whose ideal technical properties are simulated. Steve Bone explains how this can work: 'We've been working with one of our clients to set up a computer simulaton in which we can vary the way the technology works but which convinces the customer for the products that it is doing what it is supposed to do . . . The client estimates that they are about four years away from launch, but they felt that if they waited until they had prototypes, they would have invested a huge amount of money without knowing what was acceptable to their potential customers in terms of features and what was not.' Early testing can reduce risk and helps to keep the product in line with actual rather than assumed customer preferences.

Analysing customer needs

Creativity on its own does not necessarily lead to marketable products. Successful companies not only rely on traditional marketing information but also build up a bank of customer information both through customer surveys and by maintaining close contact with key customers. This kinship with customers is an important factor in the success of small- and medium-sized firms, as Dominic Swords, course director of the Part-Time MBA Programme at Henley Management College, has found. Since 1990 Henley Management College and Price Waterhouse have been researching more than 600 firms in the Thames Valley region, whose customers are located across the UK and outside it. The most recent survey, *Growth and Innovation* (Birchall et al,

1993), involved looking at attitudes towards innovation and inviting firms to discuss their experiences with Henley staff. Swords comments: 'These companies are saying to us that you have got to have in place systems by which you get to know your customers.' He adds: 'What these firms are doing is using people who have contacts with customers and then sharing the information with key figures in the organization.' This customer information can then be fed into the new product design process.

Conjoint analysis

Most large companies now use a mix of analytical techniques by which to test their creative concepts. One of these is conjoint analysis. The essence of conjoint analysis is akin to market testing: potential customers are questioned about their preferences concerning a specific set of product attributes, and they are asked to make trade-offs between different factors such as performance, design and price. The resulting information is then fed into a company's database. This technique can be valuable in refining a design. It can show how much customers would be prepared to pay for a particular product characteristic or even whether they value it at all. The information can be fed into a product demand curve. The exercise concentrates the minds of the engineers and the designers on what is actually important from a market perspective.

The success of this technique, like any other, depends on how well it is done, and especially on how good the qualitative information that it provides is. It is not the same as market research, which tends to be highly statistical. Moreover, conventional market research tends to be historical rather than forward-looking; it indicates what people have been buying and why, but it does not necessarily say very much about future trends or future preferences. This is why some leading Japanese companies claim never to use market research because they feel it has intrinsic limitations.

The value of conjoint analysis is that the information generated is focused on a product idea that the company already has in mind. The philosophy behind it is not passive as it often is in the case of traditional market research. The information that is garnered in this way can be used to practical, immediate effect.

Quality Function Deployment

Another technique used to focus on the customer is Quality Function Deployment (QFD). QFD is a much more rigorous technique and a much more time-consuming one, which is why it does not find favour with everyone. QFD matches the requirements of a product with the processes necessary to make it (such as design and manufacture). It also charts what the company's competitors are doing. This can include competitors' competences in manufacturing, distribution and selling. The assembled information can then be used as a means of ascertaining the company's weak points and the difficulties that it is likely to face. These conclusions can in turn be used for setting certain goals.

This process can become excessively bureaucratic and detached, and can be misused. Barrie Weaver of Weaver Associates cites one company that spent a year engaged on a QFD programme only to find at the end of it that it still had no guide as to what its next new product should be. QFD, like any technique, is not a magic box of tricks. Its value lies in helping a company to refine an existing product concept and in putting companies who do not communicate with their customers more in touch with them.

Larger companies tend to use QFD, but for particularly focused exercises. Many that do use it express reservations. Gordon Sked, Rover product design director explains: 'We have used QFD for fairly specific issues and we found it useful, but it is not a universally flexible tool. If you don't have the right corporate culture it will just confuse your business. QFD won't really work in the old-fashioned, classically functional environment, because you really need to have the team-based, open-cultured environment that we have built up to get the best from it.' He adds: 'QFD is a tool to help you arrive at the best design, but it won't work effectively at the macro level. You couldn't use it to design a whole car.' QFD can also prove to be too time- and resource-consuming, but can be a useful accessory tool to pure market research.

The Kenwood JE600 Juice Extractor, designed and developed by PA, matches aesthetic appeal and functionality with safety and simplicity in use. This is the key to its success: satisfying customer needs in both performance and appearance. (PA Consulting Group)

Project teams

Nearly 80 per cent of the Queen's Award-winners in the survey by Ughanwa and Baker described in Chapter 2 put the design function at 'top-management level'; nearly 60 per cent 'encouraged designers to see new products through to commercialization'; and most of the firms used a combination of engineering designers and industrial designers as part of a team to develop new products. Just over 60 per cent of these firms said that using both industrial and engineering designers 'accounted for the successful design and development of the award-winning products.' Almost all the firms – 96 per cent – said that they encouraged close interaction between R&D, production and marketing.

Multifunctional teams

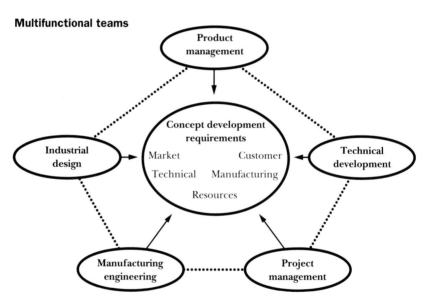

Well-managed, multifunctional teamwork is fundamental to successful product development. Input from different disciplines from the concept development stage right through the whole product development process helps to ensure that customer needs are understood and the right product developed for the right market. (Source: Arthur D Little Ltd)

It is now recognized that one way to get different disciplines to work together and to deal with design and development problems – and related management time and cost control problems – is to create a coherent product development team. Indeed, the crux of most good product design lies in team effort, but it is setting up the team and managing it which so often proves the hardest task of all. If a product is badly designed or if it is late to market, very often the reason is that there have been weaknesses in the development team.

The importance of the project team and its management is emphasized by Dr Chas Sims and Michael Cane at The Technology Partnership. They describe the problems that may arise: 'If you set up the product development on a team basis and everybody understands what the objective is and what their contribution is, then you are more likely to end up with a set of ideas and proposals that converge. The problems arise if people have an incomplete picture in the beginning. The team needs to be well briefed, but in many cases it is not. People don't really understand what is going on and what is required. Many managers think that you can parcel off the product development in little portions and make public some parts of the process while keeping others hidden on the grounds that there are some things that don't need to be widely known in the organization. But almost always it is the little things that others need to know.'

Cane points out that product development is deeply affected by what happens in the rest of the organization: 'Product development influences and is influenced by all the different parts of an organization in varying degrees, and that is a key problem; to be successful, product development has to have input from all those parts of the organization and they all have to appreciate what is going on.'

For the team effort to bear fruit 'it has to be supported from the top,' explains Arthur D Little's Chris Floyd. 'The personality of the chief executive or divisional director is one of the most overriding factors in the successful operation of a product development team. If the person at the top wants team effort, but then phones and asks why something else concerning specific project members is not working well, then that undermines the whole team, because the team members affected will dilute their contribution to the product development.'

The success of project teams also depends on functional heads having enough confidence in their role to feel relaxed about allowing teams to have control over 'their' staff. This means that the functional heads have to feel secure vis-à-vis the chief executive.

Team continuity and stability

The product development team must also be well-balanced and the type of membership consistent. This means having the right representative mixture of differing professional skills to ensure the best possible product solution, and this is why it is so valuable to bring an industrial designer into the team as early as possible.

Some managers might question the usefulness of bringing in industrial designers at an early stage when maybe the concept and even the technology is still unclear. What contribution can an industrial designer make early on? One of the advantages, explored further in the next chapter, is that outside input may challenge the assumptions of engineers or other team members and also the new product brief itself.

Many firms go wrong by failing to maintain consistent product development team membership. This may be due to lack of commitment or more likely because one or more members of the team has moved to another job in the lifetime of the product development project. A factor here is the relatively high job turnover among UK companies. For example, research carried out by Templeton College, Oxford and Mannheim University on German and UK management practices shows that in UK firms managers change jobs far more frequently than in German companies. So a UK product development team is likely to have more changes of personnel than one in Germany.

Having a consistent team can be crucial. As Michael Cane explains: 'At certain stages in the product development you are going to come up against big problems. If you've changed a team member he or she may not have the commitment because they've not been in at the start. Also, a large part of product design is learning about what you're trying to design. In the early stages you are not making much progress but you're learning. The learning aspect of the team's work is damaged if the team is broken up. A small change in personnel can make a big difference to the successful development of a product especially if time is critical.'

It is not only change of team membership but an impending change in one person's job or career that can undermine the effectiveness of that team. Many companies frequently change people's jobs. The result of this is that a person may be looking for results because of thinking about the next job, which encourages short-term attitudes. The replacement might then question everything that a predecessor has done, so generating a whole set of problems.

Visual aids and metaphors can be very valuable in helping people in an organization to understand what the next new product is aiming to be and in focusing project teams on the product and its characteristics. All industrial designers make use of small-scale models as catalysts for further ideas, but symbols, cartoons and metaphors can also stimulate the imagination. Mood boards with sets of words and images provide a means of influencing the way a product is imagined or perceived. These can work even in a traditional financially-driven culture.

David Walker of the Open University invokes the example of Japanese companies such as Honda which begin their product development with 'a powerful metaphor.' This could be a rugby player in a suit, suggesting, for example, strength and smartness. The design is extrapolated from the metaphor. This approach may not always work, but can override cultural barriers and focus the attention of team members.

Company culture

The effectiveness of a product development team is greatly dependent on a company's organizational relationships. Michael Cane argues that it is these organizational relationships that can determine how innovative a product is: 'If the company is used to a certain manufacturing technique or manufacturing a certain type of product, it will have established a whole set of pathways and ways of doing things. These may work well while the manufacturing process remains unchanged, but when the company wishes to be more innovative, it finds that it needs a lot of involvement and communication from all those concerned to achieve the necessary changes in manufacturing.' This is when major problems can arise because there is confusion about the company's aims.

The style, the thinking, the reward systems and the values of a company, which all go to make up a corporate culture, are major determinants of how well a company designs its products and gets them out into the market.

The cultural determinants of product design is a subject that has long interested James Fairhead. He argues that in most companies the prevailing management culture is oriented towards numbers, flow-charts, measurable indices and so on, and that this can inhibit good product team management. His research into the generation of new products suggests that if companies only attend to the measurable, quantifiable, officially-recorded things the chances of achieving success are diminished. As Fairhead puts it: 'People who follow the traditional precepts of project management end up getting screwed. Managing cost, time and quality is not enough. Successful project managers manage politics and culture.'

In other words, you have to manage relationships, but how you manage those relationships will be affected by the kind of corporate culture in which you operate. No matter how good the formal project management is, managers rely on informality to achieve their goals. In 1993 the Design Council researched a wide range of manufacturing companies known for best practice product development and design excellence (Farish, 1995). Almost without exception a key to their success was the ability to combine a regular reviewing and monitoring system with a flexible, conducive corporate culture. The importance of the right corporate culture is frequently emphasized by management guru Tom Peters. Its effect on design is crucial to product success.

Innovation and learning

Fairhead's particular interest is in the pre-project phase of product development. He believes that 'the effectiveness of the pre-project phase will determine the project itself' – and the pre-project phase will be influenced by the company's climate of innovation. A feature of an innovative culture is a capacity to learn from past mistakes in product development and a willingness to share information. This means not just reviewing past projects in terms of whether they met their targets but understanding what actually happened and talking about it. It means sharing information between one project and another 'as the projects go along' so that the review does not degenerate into a post-mortem, and it means analysing 'peaks and pitfalls' in the innovation process and the project development process itself. In order for this to happen, there should be people in the organization who can act as facilitators in this learning process; they themselves will be people who can think in a much more creative, less technocratic way which will encourage others to be more open.

An innovative culture is one that is both willing to share ideas and experience about the past and one that is prepared to take risks and think of new possibilities for the future. Feeling comfortable with risk is a key part of this process. The gestation of a new product is akin to scientific discovery, therefore, as David Walker of the OU explains, 'you have to be enormously tolerant of the way people explore things and you have to allow people to make mistakes, because it is by making mistakes that you learn what the constraints are.'

If this recipe sounds rather pie in the sky, it is not, for it does reflect what has happened in the case of many UK industrial inventions. The process of product development will be enhanced if the whole climate of the organization is innovative – if the organization is used to doing everyday things differently. In this way innovation becomes second nature to the organization because it is part of its thinking. The Queen's Award-winners researched by Ughanwa and Baker's were highly innovative, making constant incremental changes to their products and seeking to innovate in order to enter new market sectors. This is one of the ways in which design can be used for competitive export advantage. If the whole organization is innovative it will also mean that there can be a wider contribution to the eventual product design. This does not mean having excessively large project teams; it means that the organization has a wider pool of information and ideas to draw on.

This wider involvement of non-designers in the conceptualization of the product is advocated by Dominic Swords of Henley Management College: 'The appropriate way of seeing innovation is not to see it as something that is done by a select few. You need to have a large number of people, a critical mass of people with creative skills who can lead by example and in so doing encourage better design solutions.'

Involving suppliers

One of the factors that is changing product design is the use of suppliers. More and more companies are subcontracting large parts of their manufacture and buying in many more components than hitherto. Higher quality and environmental standards mean that companies have to be much more scrupulous in their choice of suppliers and they have to or they ought to make regular inspections of their suppliers' operations. Added to all this, product development is being speeded up so the pressure to find the right suppliers and maintain good relationships with them has become paramount.

In the motor industry, dealing with a multitude of suppliers has long been second nature: for every one person engaged in the assembly of a car there are probably 20 people making the components, and these components are manufactured across an array of countries. But now, even with comparatively simple products such as electric kettles and toasters, the ostensible manufacturer (in fact the brand producer) is likely to make no more than 25 per cent of the parts. Bill Moggridge cites an example of an IDEO-designed product – a child's guitar – that was based on input from seven locations: 'The industrial design and the engineering were done at two of our sites in the USA, the prototype tooling was carried out in Colorado, the components came from Japan and Korea, the long-term tooling came from Taiwan, the manufacture was in China and the marketing was done by Yamaha in Japan. So there you have enormous potential for things to go wrong.' However, as Moggridge points out, those kind of relationships are nothing like as complicated as they are when you have a group of companies in partnership developing a product. Then you have a mix of partners and a mix of suppliers. Managing design in this context has become much more difficult. It requires a whole new set of relationships to be established – relationships with designers in other companies and relationships with component manufacturers need to be forged. The opportunity for controlling this process in the conventional business sense is limited. Clearly, one important part of the design process therefore is to involve your suppliers at the earliest stage.

Managing the relationships with suppliers is not just a question of communication but involves a host of other factors. For example, who in the company should be responsible for negotiating the design and price of components from the suppliers? Stephen Potter of the Open University points out that traditionally this has been the responsibility of procurement people, who have a long tradition of being tough negotiators on price. The problem is that the traditional procurement culture may inhibit the inter-company cooperation that is necessary to

Involving suppliers closely in product development can reap dividends. The fact that British Eagle's Italian suppliers were able to provide them with an innovative saddle filled with silicon gel contributed to the Shadow model's Bicycle of the Year Award in 1990. (British Eagle)

facilitate successful product development and manufacture.

Another issue in managing supplier relationships is avoiding information overload. Potter cites a leading US aircraft manufacturer which has all its suppliers linked to a central database, so that the moment any design change takes place each supplier will know about it. Impressive though this may seem, in practice it has caused tremendous problems, since the flow of information is largely uncontrolled and unmanaged.

A sufficient level of information should nevertheless flow backwards and forwards, and suppliers should also be able to feed their valuable information into the product development process. The UK company British Eagle, manufacturers of mountain bikes, owe a major part of their success to close relationships with their suppliers. Close cooperation with an Italian supplier, who provided them with an innovative saddle filled with silicon gel, was a contributory factor to winning the Bicycle of the Year Award in 1990.

Just as suppliers should not be seen as detached from product development, so the management of design should not be seen as some discrete function separate from the rest of an organization's activities: product design is much too central and influential a matter to deserve that kind of limiting attention. If the design process is managed within the broader context of the organization, then it is not only likely to lead to better design but also to better performance in other areas such as procurement, manufacture and distribution. As Sir John Harvey-Jones comments: 'The creative use of design is one of the catalytic forces which combine the fields of marketing, production, research and development into a homogenous whole, which is greater than the sum of its parts.'

4 The role of industrial designers

THE MERIT OF industrial designers is that they are particularly concerned with styling, ergonomics and user-friendliness – all the factors that can make or break a product in the eyes (and hands) of the customer. Most companies, if they use an industrial designer, prefer to hire one in a consultant capacity, but some employ them full-time.

The contribution that a good industrial designer can make to a product is not well understood. Chief executives and managers are often wary of employing them, on whatever basis, because they feel that what they will get will be too fanciful or too expensive or both. The industrial designer is often seen as either somewhat wayward and anarchic or just irrelevant. But this is a caricature. Any profession encompasses a range from better to worse, but a really good industrial designer cannot only provide a creative slant on a problem but can also synthesize the various technical and market factors that impinge upon the product. As Howard Biddle of Cambridge Consultants explains: 'A very important part is played by the industrial designer, by which I mean someone who can bring together and mould the various technologies, and who is particularly good at seeing the product from the point of view of the user.'

Just as the practice of design should not be seen as an add-on to the main product development process, so too the role of industrial designer should not be viewed as some late-night extra. Although there are sometimes unavoidable reasons for bringing designers into the project at a late stage, this can mean that the way the engineering is configured leads to an almost impossible industrial design task. PA have found that this is a mistake companies make time and time again. As Michael Paton comments: 'You want to have the industrial designers working alongside the engineering and electronics people early on in order to push each other, to push the boundaries of the product and set challenges for each other, so that you draw on the best facets of all the views expressed.'

The good designer can add substantial value to a company's products, but only if he or she is appropriately employed – which means well briefed and fully involved – and only if, as with any expert input, that designer is the right person for the job. Approach and specialisms vary. Some consultancies take a more multidisciplinary approach, while others, often smaller, may be particularly skilled in product differentiation through styling.

The aim of this chapter is to show how consultant industrial designers think and work, how they form a partnership with their clients and how they create solutions for those companies. It presents the views of some of Britain's best-known designers, who have been involved in the design of some of the most successful products in the world. They have foreign as well as UK clients; indeed their overseas revenue consistently tops their income from the UK.

These designers and their clients have witnessed 'good design' work in the marketplace and have become experienced at marrying skilful engineering and electronic product design with industrial design. But what does 'good design' or indeed the word 'design' itself really mean to them? For Barrie Weaver the word design needs to be given a context to make sense: 'For us it has three aspects: appearance design – styling; development design – engineering; and mechanical design – the way it all fits together. All three are important.' Paul Priestman's consultancy, Priestman Associates, specializes in attracting customers through eye-catching styling: 'Design is about giving something character, giving it a touch of humour so that there's an element of surprise.' He cites an electric wall heater called the Cactus heater made in Continental Europe which is made of three vertical

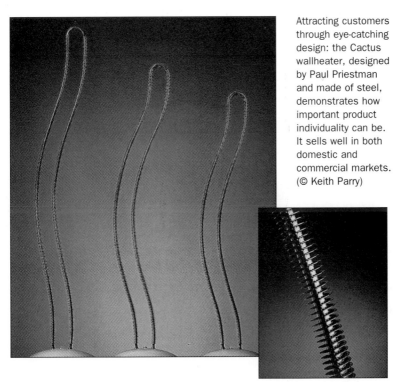

Attracting customers through eye-catching design: the Cactus wallheater, designed by Paul Priestman and made of steel, demonstrates how important product individuality can be. It sells well in both domestic and commercial markets. (© Keith Parry)

electric bars curving snake-like to a height of about five feet. It is no longer just a fire but a piece of sculpture. He explains: 'By just making a small bend in the element, you have created a whole new feeling, a whole emotion in the product at no extra cost. When people see it they immediately smile at it.'

Kenneth Grange's definition is different again: 'Good design for me is something that makes my heart lift. A product is well designed if it is a pleasure to use and a delight to behold.' Grange has helped to design most of the Kenwood kitchen equipment of the last 30 years, and has been involved in the design of razors, hi-fi equipment and the British Rail 125 Hi-Speed train. For him 'pleasure to use' takes in all the necessary performance attributes.

Kenwood Chef and Chef Excel, and the Kenwood JK700 Domestic Jug Kettle, designed by Kenneth Grange of Pentagram Design: 'A product is well designed if it is a pleasure to use and a delight to behold.' (Pentagram Design Ltd)

The designer as customer interpreter

Bill Moggridge, founder of Moggridge Associates which is now part of the international IDEO group, emphasizes that a good design is one that appeals to people: 'I always relate design to people. Therefore design is about creating products that satisfy people, that are enjoyable, that function in the right way. A good design is good for the end user.'

The ability of the designer to think into the mind of the customer is crucial in developing successful products that consistently compete well in the marketplace. Most leading designers believe that this is one of their most important roles, if not the most important role. Grange explains: 'Three parties make up the success of a new product: the designer, the maker and the user. If you focus primarily on the needs of the consumer, then the maker will also be well served and the product should then be able to sell easily. The designer stands more as a representative of the user than the maker.'

It is a role for which the designer is ideally suited by education and background. The Technology Partnership's Michael Cane, who has a background in both engineering and industrial design, explains why: 'People with an engineering or electronics education are trained in problem-solving. They are trained to make analyses by breaking a problem into bits; theirs is a world about technologies with clear-cut questions and logical answers. But the arts-school educated designers are trained to work from a different basis. They work from an emotional point of view: how will people react to this product, why would they want it, how will they use it . . . So the reason why industrial designers can be so useful in a commercial sense is that they approach things from a people perspective. It is natural to them to think about customers and markets.'

The Cellmate, a robotic machine for processing living cells cultured in sterile vessels, won an Industrial Products British Design Award in 1993. Cellmate was researched and developed by the Technology Partnership after an approach by biotechnology company Celltech; industrial design, carried out in conjunction with Weaver Associates, ensured an ergonomic frame and user-friendly interface. Cellmate created a new market worth over £2.5 million a year. (The Technology Partnership)

It is just this 'emotional' input that is often misunderstood and even feared by companies unused to working with industrial designers. Yet it is precisely this which, if properly channelled and managed, can take products over the borderline between failure and success. Cane illustrates the difference between industrial and engineering design education by recalling a question posed by a lecturer on his post-graduate industrial design course: 'If a washing machine was a person, what would it look like ?' The class then had to draw and design it. Cane, whose first degree was in engineering, says that this question provided a 'very interesting perspective on the problem of washing machine design; it made you see how products could be redesigned so that they were more fun to use.' Bridges are being formed between the two types of education all the time (Myerson, 1993), but other industrial designers also stress that their educational background predisposes them to understand customer needs particularly well.

These arguments might seem baffling and irrelevant to those chief executives who already have a good engineering designer and a good marketing manager. If the engineering designer and the marketing manager are doing their job, why would the company benefit from having an industrial designer? BIB's Nick Butler is clear

The Durabeam torch and Duracell pocket torches, designed by BIB Design Consultants. The consultancy's input brought Duracell a wide range of successful new products. (BIB Design Consultants)

about the answer: 'Industrial designers are lateral thinkers in the sense that we look for different ways of doing things. At the same time we seek to design products that are attractive, aesthetic and ergonomically sound.' In his view industrial design is the baking powder in the cake mix: 'We are an essential ingredient in the composition of a successful product. If a customer is faced with three or four television sets which are all similar, what will determine the customer's purchasing decision? They will buy the one that is easier to operate, that looks nicer. That's the baking powder.'

This is not to say that input from other disciplines is less important, nor does this conflict with the teamworking approach outlined in the previous chapter, but industrial design can often provide the competitive edge otherwise lacking in a market where competition is intense and differentiation difficult. Experienced designers such as Bill Moggridge stress the importance of the team approach in product design as does Priestman, who emphasizes: 'Design is not the most important thing in the spectrum. What is just as important is working together, especially the working together of engineers or electronics people with those in marketing.'

Nick Butler suggests that some of the UK's industries might not have declined in the last 30 years if they had had industrial designers. As an example he cites the motorcycle industry: 'The products of the motorcycle industry were designed by engineers and largely aimed at professional riders, at the people who competed in motorcycle races. They assumed that the mass of people would want what the racing drivers wanted, so that's what they designed. They forgot that there was a huge market made up of people who wanted a different kind of motorbike, that was much easier to use, for example that didn't have to be kick-started. It was the fault of the engineers and the marketing people saying that this was what the market had to have.'

The previous chapter underlined the importance of managing information and exposing designers to the information they need. Butler explains: 'Part of our contribution is that we interpret what customers' desires are, what they would like to buy. Our role is to interpret what the customer wants, not what the company thinks it should be offering, because what the company thinks it should be offering is often very different from what people will take out their wallets for.'

Although interpreting customer information is traditionally seen as the role of the marketing department, in Nick Butler's experience most marketing people are in their jobs to move products, and see their jobs as an extension of the sales function, so they look at the market from the point of view of the company's existing products. This is a view shared by many other industrial designers including Kenneth Grange, who finds that Marketing tends to be focused on the present rather than the future: 'They don't stand back from the market and ask what it needs.' A good industrial designer is more likely to question the marketing philosophy and take account of the capabilities of the engineering department.

The designer as integrator

Butler and the many other designers who share his view are not claiming that industrial designers are the fount of all good product design and development. Far from it. They understand that no single professional discipline can by itself come up with all the ideas for a new product. This is why industrial designers can be so valuable: they are used to dealing with a number of key disciplines such as engineering, marketing and production.

The idea of the industrial designer as a pivotal integrator of various professional disciplines is a very powerful one in the design profession. Those whose education achieved a satisfactory balance of engineering and arts-based subjects have a particularly good understanding of business processes and manufacturing, and appreciate what technology is all about. They are aware of manufacturing processes and engineering, and yet at the same time appreciate and are sensitive to the refinements of form and colour.

The ability to think across professional boundaries requires experience and a very client-centred way of thinking. It also means that the designer has to put aside his or her personal preferences and think in a business-oriented and practical way. As Grange says: 'It is no good imagining a product which looks beautiful but which cannot be made economically or directed at a customer who both needs it and can afford it. The designer must always be aware of wider responsibilities.'

Designers can also perform an integrating role by bringing in new ideas from the outside world: from the market, the industry or from the design world. Bill Moggridge explains: 'Someone who works in a company and has done so for a long time will know that company well and will probably be saturated with information about that company. They will be good at implementing a new product, but they may be less good at conceiving it. This may be because they are not sufficiently connected to the world with all its variety. That is what the consultant designer can provide. The consultant has helped to design lots of products for lots of clients, so there is the possibility of cross-fertilization.' Designers are therefore generalists by definition: 'We are synthesists. We take all sorts of information drawn from many kinds of expertise and out of all that we synthesize a solution. A good design will reflect those different requirements and specialisms in a well-balanced way.'

The designer as catalyst

If one talks to a wide range of industrial designers it becomes clear that, in the way they perceive their role, there are really two types. There are those who see their role as something of a magician – the inspiring, creative force behind new products; and there are those who take a more modest but more realistic view – as the catalyst or moderator of a complex set of ideas and contributions from other professionals. Barrie Weaver, who is wary of designers behaving as prima donnas – a sometimes understandable complaint from manufacturing companies – clearly takes the second view: 'I don't think you should be slightly superior and try to preach the values of design, because what the designer does has to be in tune with what the customer wants... The more companies that you see here and internationally, the more you appreciate the complexities of products, the

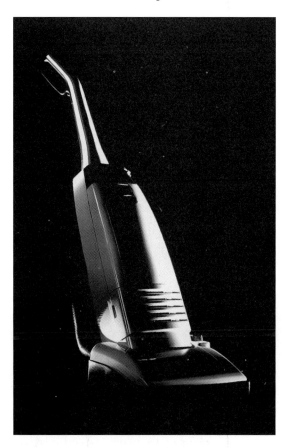

Panasonic Excel vacuum-cleaner designed by Weaver Associates: design must be in tune with what the customer wants. (Weaver Associates)

constraints and possibilities imposed by technologies, the fact that a design that appeals to a teenager might not appeal to a father and mother, or what attracts an English middle-class family might not go down well in France or Germany. Then you see the visual aspect in context, and that visual aspect is really only a small part of the whole.'

Bill Moggridge has a similar conception of the designer's role: 'What we do is only a small bit of a tiny part in a product's success, but it is nevertheless very significant. The best analogy I know of what we do is in genetics. The genes lay down the blueprint of ourselves, and thereafter it is up to us to develop the potential inherent in those genes. We can change our behaviour but we cannot change our genes. Likewise, in product development it is the five per cent or so of the early industrial design which once laid down will affect everything else that follows. So in that sense the industrial designer plays a small but critical role.'

Opening up new markets

Industrial designers can also act as catalysts by spurring companies to enter new markets. Most of the time an industrial designer will be designing a product that is essentially a replacement of an old one, but for about 10 to 20 per cent of their time they will be designing a radically new product, which may involve new uses for existing technologies. One of BIB's Japanese clients, Citizen, which is the world's second largest watch manufacturer, came to BIB with a microchip and asked what functions they would like it to perform. Butler and his colleagues were amazed by such an open-ended question. In the next few weeks, a number of ideas were suggested and rejected inside BIB. Then, on an airflight, Butler, who is a keen skier, read

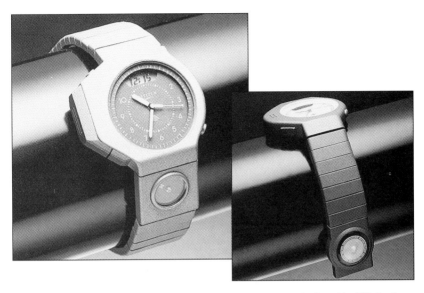

Designers can open doors to new markets. The Citizen watch, designed by BIB Design Consultants, was prompted by Citizen's asking the consultancy what functions a microchip could perform. (BIB Design Consultants)

that there were 30 million skiers in western Europe. It dawned on him that there were no watches for skiers. This sparked off a train of thought leading to the idea of a watch that was a stop-watch as well, giving barometric pressure and altitude and incorporating a compass. Butler relates: 'We went back to Citizen and asked them if their microchip could do all these things. Citizen said it could. Eventually we all agreed that we would develop a range of watches for skiers, which we did. They're selling extremely well.'

The client would not have thought of the idea themselves and in fact at first greeted it with hilarity. Industrial designers, particularly consultants, tend to see things that the company does not see. The Duracell torch (see page 69) had a similar genesis. The company asked BIB if they could come up with ideas to sell more batteries. BIB's solution was to propose a new kind of battery torch that lit up as it opened and had a beam angle that could be modified. The solution was found by analysing customer needs.

Establishing the partnership

As these examples demonstrate, when there is a good long-term relationship between designer and company a great deal can be achieved. But many companies never start that relationship at all. Most UK companies, unlike their Japanese or Continental counterparts, have no in-house industrial designers, so they do not understand the language of designers. For these companies choosing an external designer is something that they find difficult.

Sir Terence Conran strongly advises companies seeking a designer to take the time and make the effort to shop around. He is not sympathetic to those who say that they cannot work with designers, that what they get from designers is not what they want: 'If they are unhappy with their designers, then I'd say that they're not choosing the right people. I would say to them, it's your choice. It is as important to choose the right designer as it is to choose the right accountant or the right lawyer. Choose the designer who speaks the same language as you do ... Choose a designer who talks common sense but also make sure that you are talking common sense yourself.'

As outlined in Chapter 3, teamworking brings different disciplines together and helps to ensure that those involved understand and, where possible, act upon the priorities that other departments have to operate within. Michael Cane explains: 'Everybody's perspective is different and yet each one may be valid, so making sure that each person is able to contribute and shape the product is essential.' Ideally at some early point in the design process, a project team is created which brings together the consultant designers and the client's product development and marketing people. Companies and consultant designers find that this is a critical task. The relationship between the designers and everyone else 'must be

Richard Seymour of Seymour Powell believes that the relationship between designer and the client's people should be seamless. The consultancy works to make the relationship successful, as it has been at Norton. Shown here, Dale Harrow, project manager, astride one of the P55 Sports clay model bikes. (Seymour Powell)

seamless,' says Richard Seymour of the Seymour Powell design consultancy. He explains: 'We spend a lot of time making sure that relationships are right. Sometimes we act as moderators between different people in the client company, perhaps between the marketing people and the engineers.'

Barrie Weaver also emphasizes the importance of managing the interface between different professional groups including designers: 'At some stage we will have a product workshop – probably away from the client company and possibly on a weekend – in order to make everyone feel part of the design process.' He adds that it is important 'to establish a rapport with the product champions – the people who make things happen,' because at critical times it is they who will help to keep up the momentum of development.

Geoffrey Brewin (see page 118), who has observed the design process from inside a number of engineering companies, concurs with these views: 'Not only do you see conflicts between engineers and marketing people, but also between engineers and industrial designers. There are always endless arguments between these two groups, usually about design changes, either because they don't explain themselves properly or because they don't understand the impact of a particular design change on other people.'

Managing relationships is probably the most important part of product design and development. Bill Moggridge argues that project leadership is critical and that almost all the things that can go wrong on a project are to do with relationships: 'That really is the big challenge. There is a lot of talk about parallel development, multidisciplinary teams and so on, but wherever you look the theory stays well ahead of the practice.' He has a 'Big Bang theory' about why this is so. 'As life gets more complicated, the specialisms are growing apart. This means that relationships between individuals become more difficult. It means that as your products become more sophisticated, as they require more specialists, the necessity of building a team and managing the relationships in that team becomes more important.' It also means that if there are consultant designers involved they have to work very closely with the project leader.

Preparing the brief

The importance of having a good relationship with a designer is underlined in the Electrolux case study on page 118. But many design projects do not have a smooth ride. While consultant designers are used to building up a design brief many say that the initial information provided is usually insufficient. Sir Terence

Diverging disciplines? Bill Moggridge believes that while multidisciplinary working is encouraged more and more, specialisms are growing apart. ('Big Bang Theory of Design', illustrator: Philip Davies, IDEO)

Conran describes his experience: 'Frequently, amazingly frequently, in fact almost every time, somebody will come and say I need a new product and I want you to come up with ideas. But one has to say, you know your market, you have to give us the brief as to what you want to achieve.' He adds: 'Companies should show a little more responsibility and decide for themselves where they want to go. It is so frequently said, but it remains true, that a designer is only as good as the brief.'

The lack of a clear brief can be dangerous, because it gives the designer too much freedom which can prove costly and time-wasting. The problem may be caused by a misunderstanding about the way a designer works, but of course the fault is not always on the side of the client. It often depends on who in the company contracts the designer, and problems can be particularly difficult when the company is divided at board level about what it wants. Nevertheless, when things go wrong it is often the designer who is blamed. The lesson, as Conran emphasizes, is that both parties must make the effort 'to hammer out the brief so that clients get what they need.'

David Carter, Professor of Industrial Design Engineering at the Royal College of Art and Imperial College, and chairman of the DCA design consultancy, argues that it is important for the client to think through a new product design from first principles: 'Design is an intellectual process, it is not just about visual images, but an awful lot of managers don't understand this. Some companies expect you to have product designs to hand on everything under the sun. A company chairman came to me and said, 'We make lawnmowers and we're looking to develop a new model – what designs do you have?' Even if we did have a drawer full of lawnmower designs, that is not the way a good design emerges. You have to start with a blank piece of paper on which you put your marketing plans, your financial plans and so on, and out of that comes your brief for the design.'

Richard Seymour takes this point even further: 'The design brief does not start with a drawing, because that inhibits thinking. One thing I insist on from our designers is that they do not start drawing at the first meeting. Design is intellectual, it is not something that you do with a pen. The beginning of design is the sorting out of requirements. The drawing comes later on, in the middle and at the end of the design.' It is the responsibility of both sides of the client-designer partnership to spend enough time working out the brief and talking to all departments to agree objectives.

Evolving the design

Developing the design demands an understanding of two key issues: the product's market and its planned manufacturing process. The designer must understand not only the market but exactly who the product is being aimed at and their aspirations.

Early designs are likely to take shape during a series of brainstorming meetings. Whether these take place with or without the client's personnel being present varies from designer to designer. There are those who like to brainstorm a problem alone with their creative colleagues and then at a subsequent meeting discuss the possible solutions with the company, and there are others who like to work on designs with their client's people present. Most designers adopt the first course – at least in the very early stages of the design's genesis. An example is Paul Priestman, who says: 'Once it is agreed what the brief is, a group of us will meet and brainstorm the problem until we have developed a range of ideas. The aim is to allow inspirational ideas full rein and then to modify them later.

We generate a lot of enthusiasm in this process, and we try to communicate that to the client.'

Most design projects shift emphasis over time while keeping broadly to their original brief. If the project team is well managed iterative changes can be handled without too much difficulty. A key factor is how much the lead designer is in tune with what is happening to the client – in the market and organizationally. For example, is the designer aware that a competitor is about to launch a competing product and conscious that a rift has developed between the financial and marketing departments? Kenneth Grange believes that the consultant designer 'should be ahead of the client and anticipate change.' It is also up to the client's principal representative to keep the consultant informed.

A growing number of companies, however, go beyond the incremental design change and instead seek to make quite radical changes half way through a development phase. Paul Priestman has found that in bigger companies there are changes in product development all the time – changes of management and changes in company direction which all impact on the design process. A common problem is that a company will see a competitor bring out a new product, which inclines them to bring out a copycat version. When that happens companies must decide which of the two products best fits in with their overall strategy. If the original product idea answers the company's strategic needs, then it is often better to continue with it rather than change horses in mid-stream.

Creative solutions

A design concept that is merely a bit of icing on the cake is likely to bring few benefits. The more imaginative and comprehensive a design is, the more likely it is to give the product a real lift in the marketplace. Industrial design is a powerful business weapon to help companies sell their products better in both new and established markets, and should not just be used to help a company over a temporary difficulty.

Design and the management of design should be central to company activity and for any company developing or manufacturing products industrial design should be a fully integrated part of that activity. If industrial design is integral and seen to be important it is likely to be used more imaginatively and will improve performance. The case studies which follow show how this has proved to be the case in successful companies. They also show how the work of industrial designers helps companies to achieve their objectives.

5 Design in smaller companies

▶*Digital Audio Research* ▶ *The Surgical Technology Group*
▶*Solar Wings Aviation* ▶*Bluebird Toys* ▶*Conclusion*

F OR A SMALL firm struggling in the market often against
much bigger competition, good design is often the only way
that it can survive. The four small- to medium-sized companies
profiled in this chapter have all made good design a cornerstone
of their strategy. Three of them have won design awards. All of
them are high exporters, exporting between half and four fifths
of their output.

Digital Audio Research

Digital Audio Research's SoundStation was totally new in
concept, was designed and developed in record time and
launched at an international exhibition. Like the Surgical
Technology Group's Selector (see page 90), it was deliberately
designed to be easy to use – ergonomics played a big part in the
design and in its success – and it was given an excellent
appearance by a consultant designer. Another similarity is that
the new product has sold very well internationally.

Digital Audio Research (DAR) was founded in 1984 to serve
the needs of the film industry for editing audio tapes. The
founders were Jeff Bloom, Nick Rose and Guy McNally. Bloom,
who was born in the USA, had a research background in physics,

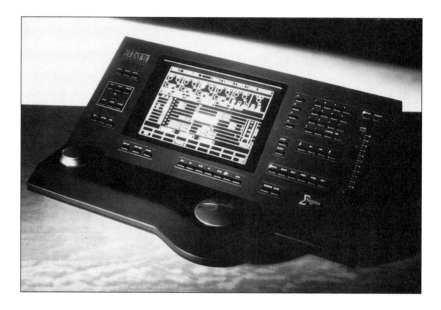

Digital Audio Research's SoundStation combines attractive appearance with thoughtful ergonomics for user-friendly operation. (Digital Audio Research)

music, acoustics and signal processing, while the other two had experience in engineering – McNally was a recording engineer with the BBC.

The SoundStation is described by the company as a multichannel, disk-based digital audio recorder, editor and processing system. It is used in the production and assembly of edited multitrack audio material and provides a relatively easy way of manipulating and editing sound for audio tapes in film, television and video applications. It was the company's second product.

SoundStation originated from discussions that Bloom, Rose and McNally had been having in the early 1980s. They were all interested in digital technology, especially in the film industry. Around 1981, says Bloom, 'we stumbled across the fact that dialogue synchronization in the film industry was done in an incredibly archaic way, and it encouraged us to see whether it could be done digitally, which would speed the whole process up.' Bloom went off to simulate a computer programme that would automatically synchronize dialogue. The result of his endeavours led to DAR's first product, which was called WordFit. This was sold to a number of film studios, mostly in the USA, in 1985. By now DAR had a fourth member, Mike Parker.

Although WordFit was a success, DAR recognized that it would always have a limited market. It was expensive and it was designed for a very specific purpose – synchronizing and editing dialogue. So the DAR team set to work to develop a more general-purpose editing system. This became known as SoundStation. In order to develop this, DAR tried to attract venture capital, but eventually, in 1986, decided to accept an offer from Carlton Communications to buy the whole company.

SoundStation was envisaged as being totally different from any of the competitive products. Bloom explains: 'We set out to make a system that would enable a sound editor to use it in an intuitive way, that would be easy and simple to use. The systems that were then on offer had much more of a computer interface, which meant that if you wanted to go through a particular operation you might have to press a keyboard button, which then gave you a menu of alternatives, which in turn would then mean that you had to press another button and so on. So you could be pressing ten buttons or more to do something which on the SoundStation requires only three. On the SoundStation all of the commands that you use are either hard buttons that are always there or touch buttons on the screen that do not disappear. So the screen stays pretty much as it is throughout the whole editing operation.'

The SoundStation also has a novel sound segmenting

technique. Any piece or segment of a recording can be isolated and numbered then modified as required; the sound editor can see these segments on the editing screen. But the original version of the segment stays the same and the editor can switch back to this at a touch of a button. The system has a number of editing and processing tools which include cutting, moving, naming, trimming and repeating segments.

The project required three major areas of design: hardware, software and interface with the user. The hardware and software were designed by DAR's three founders and Mike Parker, with some help from other electronics engineers who were recruited. The interface was largely designed by Jeff Bloom and Nick Oakley, a consultant designer who was then at the consultancy Seymour Powell.

DAR called in their design consultant very early on in the design phase. Bloom says that the company had to do this because the design team were determined to have SoundStation ready for a US engineering exhibition and conference in November 1987. The company also brought in design consultants early to design their corporate literature and logo. Many companies do this much later.

The interface of the SoundStation became a key selling point. It was not simply that it looked exceedingly stylish or that the angle of the screen was right but also that it was highly tactile. For example, the main right hand control was made of solid brass rather than from other material such as plastic or aluminium. The result, says Bloom, is that 'you feel a solid connection between yourself and what you're turning. There's no inertia in the control which is what you often get when it is made of other materials, so your ability to modulate the sound and edit it is enhanced.' The fact that the controls are much easier to use and much more self-explanatory means that a sound editor can concentrate on the main creative task in hand rather than be distracted by the operational mechanics of sound-mixing and editing.

When the SoundStation was unveiled it was a huge success. The fact that it was hugely photogenic and performed superbly lay behind the amazing response. Bloom comments: 'I've always been a believer that the appearance of something is incredibly important especially in computer devices where so many things look the same, but this success made me realize this even more.'

Three lessons came out of this design process. Firstly, DAR found that 'everything takes longer than you think.' Although the product was far enough advanced to launch it by the due date, in fact another year was required to 'hone the software and make it absolutely reliable.' Secondly, Bloom and his colleagues appreciated how important good design was. He explains: 'We knew that already, but the reception that the product got confirmed it.' Thirdly, they recognized the value of good ergonomic design. 'People could see that there was a big difference between using SoundStation and anything else. This product freed their creative abilities.'

One advantage was that the design was modular-based. This was done deliberately so that it could be developed further in the future, which has indeed happened. There are now several variations on the original model aimed at specific customer groups. The SoundStation's visual and ergonomic design combine to delight and empower the user.

The Surgical Technology Group

The Surgical Technology Group Ltd has seen its turnover rise by more than 20 per cent a year since it launched a product called

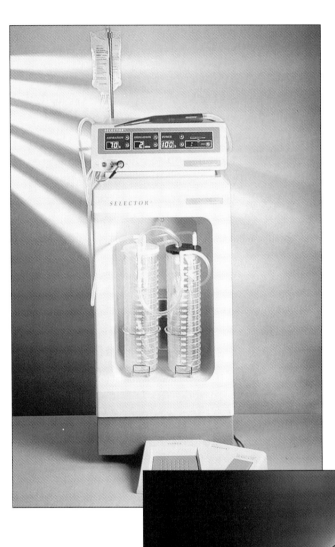

The success of the Selector ultrasonic surgical aspirator is due to careful market and user research, and the careful fusion of engineering and industrial design skills. Detail shows Cuschieri laparoscopic handpiece. (The Surgical Technology Group)

Selector in 1990. Selector is an ultrasonic surgical aspirator. It is essentially a high frequency 'hammer', which destroys body tissue such as tumours, by subjecting it to intense vibration while leaving the fibrous material, such as blood vessels, unharmed.

Developing this product was a high risk strategy on two counts. The company had recently been the subject of a management buy-out (in 1987) which meant that if the product failed then the directors would be personally liable – the development costs of £400,000 were high for a company with a turnover of less than £2 million, and the group had no experience in the ultrasonic field. Yet the aim was to bring out a product that was not only technically competent but streets ahead of the competition. On top of that the company's design capability had been run down by the previous owners – there had been four over a fairly short space of time – so effectively there was no research and development facility and no serious product development. So what was the thinking behind this product and why did it succeed?

Peter Gibson, managing director, explains: 'The business we took over was profitable but its sales were constant, and as the cost base rose our profitability was declining. The product range was limited to one technology – cryosurgery, which is used for freezing tissue and destroying it. We had pioneered this. We had to find another product line, another technology for tissue destruction.'

Gibson, who had a background in clinical biochemistry research and latterly in the design of analytical instruments, saw several possibilities. The company could develop a product in one of three technologies: surgical lasers, electro-surgical equipment or ultrasonic equipment. He and his co-directors were most attracted by ultrasonics. The company felt that ultrasound presented many opportunities because it wasn't being used very much as a tissue destroyer.

Examination of the market showed that existing products were large, expensive and complex to operate. Peter Gibson explains: 'We felt that the main competitive product was not well designed. It took a nurse 30 minutes to put the instrument together and it had too many controls, some of which were not easy to operate. There was another problem with it: to sterilize it you couldn't put it in a steam autoclave unit, you had to immerse it in sterilizing liquid and that liquid sometimes caused skin reactions.' Research also revealed that the original patents on the basic technology were about to lapse.

The brief for the new product began to emerge. The Surgical Technology Group wanted to make a product which did not need to be put together by the user, which had far fewer and simpler controls and which could be sterilized more easily – but the company still did not have the expertise to develop ultrasound equipment. Peter Gibson is the first to admit that the move into ultrasound was very ambitious. However, he says: 'What we did have was some good core skills such as fine mechanical assembly, experience of using materials that could withstand autoclaving temperatures and a good manufacturing capability.' For the actual ultrasound technology, Spembly took on the services of a technical consultancy.

Before the technical consultants came in, there were three important developments. The Surgical Technology Group sold off one of the other companies in the group in order to raise the product development finance which it needed; three young engineers were hired to beef up design capability; and the company began to sell a related Japanese ultrasound product as a preparation for entering the market with its own model later.

The design of the Selector was clearly focused on the customer. Gibson talked to a number of surgeons about the existing ultrasound tissue destruction equipment and also about what they and their nursing colleagues would prefer. Among a carefully chosen few he discussed his company's own product plans. The essence of the new design was going to be that it would look attractive and would function well. 'Having an attractive design,' was very important to the product strategy. Indeed, Peter Gibson believes that much of the commercial success of the product has been due to its good looks.

Interestingly, the company only brought in industrial design consultants towards the end of the whole design process. This was partly because at the early stage they were still deciding what they were trying to do, so it would have been difficult to have kept the continuity if the industrial designers had been involved from the beginning. When they were brought in the relationship worked well: 'What we did do was explain to the consultant designers how we'd got to the point where we were, how we'd arrived at the final product solution. We took them through the learning curve pretty fast. As part of that process, we spent a lot of time with them explaining what the product would be used for, showing them surgical videos, for example, so they fully understood what the product would be doing and the context in which it would be used.'

What the consultant designers saw when they visited the company was an engineering model. This had been developed by a project team composed of Gibson and the three specially recruited engineers. As the product design matured, the team members would discuss it with engineering technicians and assembly technicians in the company. They also met the technical consultants at intervals. Throughout the whole product design phase Gibson would be 'thinking of the customer's requirements and keeping in mind a vision that we could come up with a radically different design, though not knowing whether we could or not.' He admits that for a while the development seemed to stumble, but that it was principally the consultants and in-house engineers who eventually said that what they were trying to do was feasible and achievable with their resources.

The biggest problems were time and money – the same problem. As Gibson explains: 'The engineers wanted more time, but we didn't have the time because we didn't have the financial resources. We were spending all the cash we had, so there was a very strong commercial pressure to do it by the time we'd set ourselves.' The financial constraints also meant that the product's manufacture had to be costed very closely and the cost parameters had to be kept to. Gibson says: 'We defined the production cost very dramatically and firmly at the outset, and we made people watch the cost of manufacture every month. We overshot it only slightly.'

In the latter stages of the design, the whole team – which by now included the consultant designers – was working through the night every night till it was finished. 'Miraculously, we got it out in time.' The timing was important not only for financial reasons but also because the Selector was going to be launched at a US medical equipment exhibition. The USA is the world's biggest medical market, so if the product missed its US launch that would have set back the sales considerably.

Towards the end of the development phase, the group's institutional backers, seeing cash draining out of the company, got worried. Peter Gibson explains: 'We'd told them at the outset that there was a risk involved in what we were doing, but I don't think they realized that we meant it. When we had tripled our profits and were set for growth, the backers pulled out.'

The product was a success, because it combined good design with good ergonomics and unlike its main competitor it had six controls not twenty-six. What was also important was that it was heavily marketed overseas. Within eight months of the launch it was selling in twenty-three countries. Gibson comments: 'All through the development phase, whenever I happened to be abroad I was continually looking at other products,' so he was mentally prepared for selling the Selector internationally once it came out.

One of the reasons why the company was able to develop this product in a fairly short time – 18 months – was that it had a small project team and no communication problems; Peter Gibson was research and development manager and marketing manager at the same time. Today, both jobs have been delegated as the company has grown so fast. So the challenge now is trying to work in small teams again and maintaining close communication between the marketing and the R&D departments. One way to do this, says Gibson, is to 'make the product manager live in the R&D department for a while so that they all see that the point of what they do is serving the customer.'

The Selector was based on understanding and serving the customer well, and as a result it beats competitors.

Solar Wings Aviation

Solar Wings is a small microlight aircraft manufacturer, which by dint of clever engineering design and first-rate industrial design has markedly strengthened its position in the international microlight market. Roughly half its production of 150 microlight aircraft in 1994 was exported, whereas in 1988 only about 20 per cent went to overseas customers. This shift in customer base is directly attributable to a change in design policy.

Solar Wings was set up in 1979 by microlight enthusiasts. The products were acceptable by industry standards but pretty basic and flimsy. The seating was also cramped; the navigator and the accompanying passenger in this and other microlights were not expected to be up in the air for very long. Dr Willy Brooks, who was the company engineer until the spring of 1994, explains that to some people the company's early products looked like 'motorized deck chairs.' In short, the company was making bottom of the market products.

In 1987 Solar Wings decided that it wanted to build a rather better model as part of a strategy to move the company and its products up-market. Dr Brooks, who had carried out a considerable amount of airworthy certification work for the Civil Aviation Authority and who had recently completed a PhD in aircraft design at Cranfield Institute of Technology, joined the company at about the same time. The new product brief for Dr Brooks and John Fack, marketing manager, was for a microlight that would look less functional and which would inspire more confidence. As John Fack explains: 'We wanted something that would be more comfortable, would have better aerodynamics and which would have greater stability and suspension. It would look more like a serious aircraft rather than a toy.'

Fack himself was part of the flying public. He knew from his own experience and from what other enthusiasts were saying about Solar Wings' products and those of other companies, that people wanted something much better from the microlight industry. They wanted comfort, reliability, better manoeuvrability, a less shaky ride and less noise.

The new model would use a Rotax engine, as before. The main difference would lie in the purpose-built seats, replacing

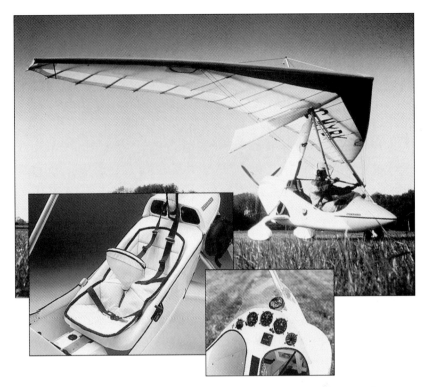

The Quasar Microlight differed from previous models in introducing purpose-built seats, a solid one-piece frame and restyled undercarriage. A marriage of clever engineering and industrial design achieved better comfort, reliability and manoeuvrability. (Solar Wings Aviation)

the old canvas ones; a completely solid one-piece frame, replacing the previous mix of tubes, aluminium and canvas; and a restyled undercarriage.

Brooks' drawings progressed rapidly towards a rough prototype in which the seats were cast from glass fibre while the undercarriage was made from glass epoxy fibre. The undercarriage was built in one piece and flared to give better aerodynamic properties. The basic structure was made from aluminium.

This first model was then taken through a series of changes in order to improve its steering and lower its weight. One of the initial problems with the much more ambitious design was that rendering it stronger also made it heavier. The parts of the design that used composite materials had to be reworked several times. In addition, the more sculptural appearance meant that the aircraft's aerodynamic behaviour was different.

Quite early on the company brought in Seymour Powell as design consultants. Seymour Powell's design work was largely paid for by a DTI enterprise grant. The consultancy played a key part in styling the airframe. A smart part of the eventual design is that the undercarriage almost merges into the frame, which makes the whole aircraft look much more solid. The previous model had an undercarriage that looked as if it had just been screwed on like an extra piece of metal.

This new aircraft, called the Quasar, transformed the company and the styling was clearly a selling factor. As Fack says: 'It gave us a lever into markets that we had never been in before. By having a product that was startlingly different and beautiful we were suddenly no longer a little manufacturer that was more or less bound to the UK home market but we were in demand all over the world. The photograph of the Quasar alone got us into some export markets.'

Since the arrival of the Quasar, the company has gone on to build the Quantum which is a derivative of it. This is a cheaper version, which can be adapted to specific customer needs. For example, game park wardens in Africa have modified the Quantum for surveying the wildlife, Laplanders have fitted skis to it and shoreline patrols in South America have adapted it for their needs as well.

John Fack has seen the benefits of good design for himself while travelling abroad selling the company's range of microlights: 'I think we appreciate the need for good design more than ever. As the international market gets more competitive the models that are better designed will outlast the other ones, especially if the design not only gives you a product that looks better and flies better but is also easier to produce.'

Bluebird Toys

Bluebird Toys was one of the entrepreneurial success stories of the mid-1980s. Founded in 1980 by Torquil Norman, who had previously been managing director of the British toy company Berwick Timpo, the company attracted attention early on with its Big Yellow Teapot. The inspiration for that product came from Norman who designed it himself. His aims were to create well-designed toys for young children, to produce them at a good profit and to place the manufacture of those toys in Great Britain, thereby keeping overheads low. By 1986, the company's sales were nearly £12 million and pre-tax profits were £1.7 million.

By 1990 the situation changed, the company's growth combined with the recession was causing financial problems and the company went into loss. Following a fundamental restructuring which involved closing three factories and moving a substantial part of the production to the Far East, Bluebird's managers are now confident that they have re-established the basis for further growth. In 1993 sales were nearly £69 million and pre-tax profits were £9.8 million. In the same year 56 per cent of sales went overseas. In 1994 sales were nearer £100 million with 65 per cent of profits from overseas.

From a design perspective, Bluebird Toys is interesting for a number of reasons. It is operating in a very volatile or choosey market; it is launching new products at least every three months; and it has long lines of communication with several of its manufacturing suppliers and with some of its product designers. So there is considerable potential for getting things wrong. The product designs come from three sources. Some products come from professional toy inventors; these are mainly drawn from the UK, the USA and Japan. In such cases the manufacturer will develop and produce their toys under licence. The Mighty Max toy, which is a product that opens up to reveal a miniaturized world, originated from inventors in Chicago. Bluebird Toys bought the concept and then developed it and named it Mighty Max. Between 20 per cent and 30 per cent of Bluebird's toys come from external sources.

A second source is Bluebird's own designers. They were the source of the Big Red Fun Bus and products in the company's craft range. A third source, which has proved very successful, is commercial toy designers. The company works closely with one such firm in London and another in Chicago. With the toy designers, Bluebird will often take an idea to them and ask them to develop it. The Lucy Locket's Fabulous Dream Home toy for girls which opens up to reveal a miniature doll's house was developed by the company's consultant toy designers in London. More than half the company's sales now derive from their toy design consultants in London or Chicago. The advantage of using these outside designers, says Casey Norman, who manages Bluebird's product development, is that the company has built up excellent working relationships with them so the designers will know if an idea of theirs will be attractive to Bluebird. If the idea comes from Bluebird itself they will understand how the product would be manufactured and where it would be sold.

Ideas for new products are assessed at product meetings which are held monthly – by toy industry standards very frequent, but this means that the company can react to inventors' ideas

Bluebird's Big Red Fun Bus: with different product design sources, the company concentrates on effective communication and project management and aims to appeal both children and parents. (Take 2/The Design Council)

and to the market quickly. Those present at such product meetings include the chairman (Torquil Norman), the chief executive, the sales director, the development director, the international sales director, the technical director, whose additional brief is to supervise manufacturing, and two people from the marketing department.

The product appraisal involves detailed costing which is established at the meeting. This includes: sales price, volume of sales expected, especially in the main markets, the cost of developing an engineering model, the cost of tooling and packaging, the royalty to be paid where applicable and the cost of advertising and promotion. Agreement in principle is reached at the meeting, and all the financial information is given later to the financial director for checking through.

A great deal of attention is given to the styling and packaging of the toys. 'Getting the look right is an absolutely key consideration,' says Casey Norman. 'The products have to be fashionable in their design and their colours. If a child feels that the toy is a bit passé, just a tiny bit out of date, the news of their disappointment will travel fast. The wording on packaging is also important. A badly chosen word can put a child or a parent off, so you have to keep in mind those two groups of customers.'

One key to the company's current success has been strong project management. It is the management process driving

Bluebird's Lucy Locket's Fabulous Dream Home was developed by the company's consultant toy designers in London. It is one of the products which contributed to Bluebird's huge sales growth since 1992. (Take 2/The Design Council)

products from concept to launch that has seen the greatest change since 1990. Every new product is now given a project manager from day one of the decision to go ahead with it. As Casey Norman explains: 'They will be responsible for the progression of the product right to the end. This does not mean that they will make all the decisions. They might not decide on the name of the product or where the model should be made and they might not be involved in negotiating the contract price with the manufacturer – they might draw on the expertise of people in other departments. But they will be responsible for the overall movement of the product towards its launch.'

The practice of using the skills and experience of others is a fairly recent development, and it is one that has been forced on the company by its huge sales growth since 1992. (Group turnover rose by 50 per cent in 1993, while international sales increased by 142 per cent). Whereas project managers used to do everything, they now act as shepherds keeping a watchful eye on a number of different product lines.

Of course this change, whereby one person is managing several projects at the same time, each of which will involve other groups and specialists, brings its own problems. Some toys have their design changed as they go through stages of development, often because of changes in the market. In such cases it is absolutely essential that everybody concerned knows about the change of design. As Casey Norman explains, the challenge is 'keeping the flow of communication running smoothly.' The company has quite consciously set up mechanisms to ensure this. These include the appointment of an internal co-ordinator purely to facilitate communication and Casey Norman's own biweekly meetings with all the project managers and all the packagers. At these meetings they look at their current and future products. Using critical path analysis they monitor the schedule for each one.

Conclusion

Bluebird Toys is the largest of the four companies featured in this chapter and by some definitions it is no longer a small company. The day-to-day problems that it faces are to a large extent different from those of DAR, Solar Wings and the Surgical Technology Group but like the latter it has grown very fast. For both companies project management has become the most important issue. What the four companies share is a drive towards good design, based on good market knowledge that is supported by frequent trips abroad. It is quite clear that these four companies do not use market research in a passive way to aid design. All of them actively look at their markets for inspiration, for ideas about customer needs, and in each case this means travelling outside the UK to see what people are buying and what the competitors are selling. They all test their products on customers and they have all matched good design with an understanding of their customer.

6 Design in larger companies

IN LARGE INTERNATIONAL companies there are usually two design-related problems. Either the design function is lost in the mayhem of the corporate structure so that it is sidelined, or it is badly managed whether the structure is federalist or matrix driven. These two issues – making design central to the whole enterprise and achieving a coherent way of managing the design process in the context of the organization as a whole – are of immense importance to international companies. They need to be handled well if the company is to react effectively to the many pressures acting upon it.

The managers of medium-sized and small firms often think that the problems of these large international companies are not relevant to them, and they may also feel that large companies are much better resourced so their solutions are not relevant either. But this is a mistaken view. The difficulties experienced by global companies are to a large extent those of the smaller company writ large. They are essentially about responding imaginatively to the market and getting people to work together. If the big company has the advantage of access to more capital and the opportunity to tap into the expertise of specialists, it also has one major disadvantage which is that it always finds it more difficult to change.

The following examples show how four companies have

managed the process of change, principally change in design management.

The Rover Group

The Rover Group transformed itself between the end of the 1980s and the early 1990s to such an extent that during the 1990/1993 recession it increased its share of the UK and the Continental car markets. Part of the reason lay in its new approach to design. However, there were other reasons, the two key ones being a complete change of company culture towards a team-based business philosophy and new manufacturing processes. These three factors (design, culture and manufacturing capability) were all linked.

If you ask anyone in Rover today how the company approach to design changed between 1988 and 1994 they will mention four things: an accent on teamwork, the widespread use of Simultaneous Engineering (often described as Concurrent Engineering), a greater emphasis on pre-concept design work and deeper market research. These are all identified by Gordon Sked, Rover Group product design director, as having been of major importance, but he maintains that there was one key initiative that predates all of that and without which these four improvements could not have happened. That is quality improvement.

The Rover Group has launched three quality initiatives since 1987, each one building on the previous one. The first in 1987 was Total Quality Improvement (TQI). This set out the concept and spread the message of Total Quality. Beginning at the top of the company, TQI encouraged everyone from directors to dealers to talk about quality and improve their own working methods. In 1988 Rover adopted Total Quality Management

A new approach to design and quality management, supported by changes in corporate culture, has brought Rover and its products dramatic improvements in performance. Products and people are more important than particular functions. As Richard Elsy, product director for Land Rover, confirms: 'Everybody's energy goes into a product, and everybody wins.' (Rover Group)

(TQM), which established clear techniques for managing quality. Then in 1990 the company launched Total Quality Leadership (TQL), which was seen as more an inspirational tool for really involving people in quality improvement. Sked says that for Rover at any rate there has been a significant difference between TQM and TQL. 'We found the concept of quality leadership a much more powerful one than quality management; it reflected our culture much better.'

The drive for better and better quality remains an axiom of Rover. Everybody has two tasks – their day-to-day job and improving the way they carry out that work. Employees are expected to log their improvements in a book that they keep for this purpose. 'That is very valuable,' says Sked, 'because the process of writing down what you do makes you review it more carefully.' So the notion of continuous improvement is in people's minds all the time.

More importantly, the accent on quality became 'the oil in the corporate culture,' and effectively helped to create a greater degree of consensus in the company. It gave the company a common language. With that foundation the Rover Group went on to make even more significant changes. These changes had a critical impact on product design.

The first of these was setting up dedicated project teams. Richard Elsy, product director, Land Rover explains: 'In the culture of the company today, the function – be it engineering or marketing or finance – is no longer the important thing. What is important are products and people. Whereas people used to get their kudos from being part of a function, now they are stimulated by being part of a project team. Everybody's energy goes into a product, and everybody wins.'

Elsy points out that these project teams are not engineering-based teams. Membership is much wider than that. On his project teams he will have people from engineering, design, manufacturing, purchasing, marketing and finance. The members of the teams are expected to take a business view of the product and not just a functional one.

These teams, says Sked, are a microcosm of the company as a whole; the practice of having an array of teams working at any one time means that a business philosophy pervades the whole organization. Sked sees a noticeable difference in the company as a result. He says: 'We all used to talk in a much more functional way, but now we talk much more in business terms.' He adds: 'Having a business view means that you have an open mind. You're not pushing your own viewpoint, and that is a hell of a difference.'

The second big change was Simultaneous Engineering. The essence of Simultaneous Engineering, which is applied principally to new product design and development, is sorting out a range of key problems all at the same time. In practice it also means dealing with problems as early as possible. The thinking behind this approach is that the more effort and the more resources you put into the early stage of a product development the more you will save later in terms of money and time. If you 'front load' your problems and find out what are the tricky things and solve them early on you avoid the sort of problems that traditionally hold up the delivery of your product. So this means setting out in detail what components will cost, what the manufacturing processes and costs will be, what are the special features that the marketing people need. It means talking with suppliers very early on too. Therefore, instead of dealing with the separate parts of the design and development process in little pieces and sequentially, as much of the whole process as possible is discussed simultaneously. Indeed Elsy talks of sorting out many of these problems 'even before the product is designed.' The result of these intensive discussions should be that 'when the design is released and the component parts are procured, those parts can be fitted straight on to the car. You have no re-engineering and no re-design.'

The first time that Simultaneous Engineering was adopted was for the Land Rover Discovery (see page 107). By using this system Rover was able to move from the product approval stage to volume production in 28 months, at a time when the best industry standard in western countries was 36 months.

Although the Discovery project was taken forward by one team, the members of that team relied for support from others in the organization. As Elsy explains: 'All team members had

Rover's recent car designs are based on deeper and more exact customer research which looks at both 'hard' functional and physical criteria and 'soft' issues such as the product's personality. (Rover Group)

responsibility to progressively cascade out involvement to the rest of the company. The capability to hand over to the mainstream disciplines in an orderly way was an important element in the project's implementation.' He adds: 'The location of team members cannot be underestimated as a key to success. It may give rise to a number of practical and logistical problems, but at the end of the day there is no magic formula. A bulk of Simultaneous Engineering is about effective and realistic face-to-face communication.'

The combination of Simultaneous Engineering and dedicated project teamworking on the Discovery was not achieved without pain. This was the first exercise of its kind in Rover. With hindsight the company realizes that it should have prepared the organization for this radical step. Elsy, who managed the Discovery programme, says that it 'rocked so many barriers in the organization.' Not only did the programme work across so many functions, but it also made great demands in its early stages on certain people in those functions. As Elsy says: 'Project teamworking gets very resource intensive because it can tie up people for long periods of time.' Although the programme was a big shock to the Rover organization, the benefits were so obvious that the company decided to restructure the whole company along project lines.

As Simultaneous Engineering has progressed throughout Rover, the company has learned more about how to manage the process. An important point is to control the amount of change within the project. Checkpoints or milestones are put in place after which the freedom to make alterations to the design or development specifications is restricted. Everyone on the team is aware of these control points. This is not as rigid as it may sound. 'You can still be flexible', says Elsy. 'If, for example, an engineer feels that they need an extra two weeks to work on some part of the design, they can usually trade that off against something else which might be delivered early'.

As the process has matured within Rover, so its senior managers have been trying 'to look at ways of pooling resources back to the functions a little, making the project teams smaller but sharing resources much more, so that people can work on several projects more easily.'

A third change in design philosophy is strengthening the resources that go into the pre-concept stage of the product development. The pre-concept stage is the period when ideas for new designs and products are thrown around. Discussions are open-ended and much less formal. By the end of the pre-concept stage, managers will have a fairly clear idea of the new product's shape, size, market sector and so on, so that in the concept phase the managers are tackling something specific. Sked believes that if you do not put enough effort into the more fluid pre-concept stage, then you are limiting the potential of the concept itself. Although people do not come to the pre-concept stage with fixed notions of what they would like and what they feel the company needs, nevertheless quite a lot of definitive information is brought to bear on the discussions. This includes technical, marketing and customer information.

A fourth change has been greater attention to market research, or to be more exact customer research. This research can be very effective at the pre-concept stage. The company has evolved a technique for assessing customers' preferences about existing and proposed products. The research looks at 'hard' characteristics such as performance and physical criteria and 'soft' issues such as a product's personality. What emerges, says Sked, is 'a whole series of answers against which we can test some of our vehicle concepts. It's an interactive process that gives us the opportunity at the very earliest stage to say whether one out of several concepts is more customer-targeted than another. It is still a rough tool, but it is much better than anything we had before.'

Rover's advice on Simultaneous Engineering

- Do a pilot project first, but one that is meaningful, that is, something that you need to do.
- There will be turmoil in the organization and you will make mistakes. You deal with the turmoil by counselling and coaching the functional heads.
- The task should be something that you have no idea how you would achieve using the conventional approach.
- Lead from the top.
- Pick the most imaginative people, not necessarily the most skilled but those who are not traditional functionalists, people who are a bit of a maverick and also who are energetic.
- Empower the team. Give them money and put them in control of their own budgets.
- Give them measurable targets by which to judge their performance.
- Support them within the organization.
- Deploy your vision of the future. Don't keep the project secret and explain the unarguable benefits. It does have problems but every other way has more problems.

Black & Decker

One of the striking features of the design function at Black & Decker's UK manufacturing centre in Spennymoor is how it is centrally placed on the site. The reason for this is to bring design and manufacturing closer together in every sense to improve communication and co-operation.

Another striking feature is that Black & Decker has its own industrial design group. It does use outside consultant industrial designers but only occasionally. Laurie Cunningham, who manages Black & Decker's industrial design department, is naturally a strong advocate of the discipline. He says: 'The value of industrial design is that it is such a differentiator. The competition in the consumer goods market is increasing rapidly as more countries enter this sector. A good industrial design is what can give your products that edge. The contribution of industrial designers is that they can stand outside conventional thinking. They look at things from a different angle and this difference of approach can lead to innovation in design.'

Like the Rover Group, Black & Decker has taken the road towards more teamworking and has at the same time been restructuring its organization. This large international company, with factories strung across several continents, has moved towards a matrix-style organization since 1990. It is now structured around several product group areas such as cordless products, drills and outdoor products. The functions of engineering and marketing remain, but these are seen more and more as resources supplying the needs of the product business groups.

The practical effect in a site such as Spennymoor is that there is a much less hierarchical organization. Decisions have been pushed downwards so that staff such as Peter Craddock, design

Black & Decker
uses industrial
design as a
differentiator.
It encourages
close cooperation
between design
and manufacturing
and ensures that
training helps staff
to adopt a team-
based approach.
(Black & Decker)

and development controller, see their job 'more as a mentor facilitating the work of others.' The teams, he says, have become very strong. Senior management does stand back. As part of this team-based approach people are encouraged and receive allowances for training so that they are multiskilled.

This team approach has enabled the design process to become 'holistic,' says Cunningham. There is a shared responsibility for the development of new products. The brief for a new product will be drawn up by a team of people representing industrial design, engineering design, marketing, manufacturing, manufacturing engineering and purchasing. Suppliers may also attend some of the early meetings. The Spennymoor factory, which is one of the largest consumer products centres in Europe producing about 11 million units across a 100 products or more, sources most of its components from outside, so the role of suppliers in design is important. The factory is 'talking to its suppliers earlier and earlier in the design process.'

At the same time, the whole design process has been reviewed in order to deliver new products better and faster. The multi-staged design process has been streamlined into fewer stages, thereby shortening the time needed to complete the design. One element in this has been the acquisition of advanced CAD modelling systems. They are, says Cunningham, 'more designer-friendly and more interactive, and they save time.' The CAD modelling systems are used very early on in the evaluation of new products.

These early models are used in tandem with more advanced market research and Quality Function Deployment (QFD). In other words, the design specifications are built up from three sources: industrial design, market research and QFD. The more personal, more creative input from industrial design is balanced against the 'foundation of customer expectations,' and all input is tested against information derived from the more rigorous QFD technique, which will include an analysis of competitors' products and aims.

The management at Black & Decker does not use QFD exhaustively because the technique is very time-consuming, and in any case some kinds of information can be obtained in other ways. Its main value is that it enables the company to prioritize what new features will be the most important – which ones will give products the best competitive edge.

A few years ago it was suggested at Black & Decker that the company ought to have a well-specified rotary mower to complete its range of electric garden equipment. This idea was considered to be rather ambitious since the rotary market was very crowded. However, the industrial design department, supported by the engineering department, eventually came up with a design, costed for manufacture and sales margin, which incorporated a number of innovative features: the grass bag could be emptied easily and completely and its capacity was greater; the load-bearing wheels were wider in order to create fewer ruts in the earth when the grass was soft or wet; the wheels could be adjusted to vary the height of the machine; the product as a whole had a more high-tech appearance. The model was launched in 1991 as a top-of-the-range product. Although Flymo is still the UK market leader, the Black & Decker product has carved a niche for itself and is profitable.

Laurie Cunningham comments: 'One function of designers is to stimulate new ways of thinking. The designers and the engineers were able to take models and drawings and say "wouldn't it be exciting if we could make a mower that looked like this". With a model you can capture the imagination.' He adds that what they presented was not just an attractive product but something carefully costed so a business case could be advanced for it.

In 1993 one of the company's new products won a Design Council Design Award. This was the improved Dustbuster/Spillbuster, a cordless, rechargeable hand-held vacuum cleaner, which was designed to be used in confined spaces such as underneath furniture and in corners. The product's performance was much better than its predecessor, in fact 30 per cent more efficient. It also had a removable rechargeable cell pack which could be returned to the company's service centres.

One of the new vacuum cleaner's special design features was

The performance of Black & Decker's Dustbuster range is 30 per cent more efficient, while the design of the range makes it considerably easier to manufacture and service. (Black & Decker)

its substrate, which eliminated all internal wiring, allowing much greater automation of the manufacturing process. It was the redesign of the fan and the internal contouring coupled with higher motor efficiency and better battery cell design that produced the increase in power per cell.

For Black & Decker skilful industrial and engineering design input into a well-managed development process brings its products a powerful competitive edge.

Electrolux Floorcare

Electrolux Floorcare, which is part of the Swedish international group Electrolux, has also been developing its design philosophy over the last ten years. Geoffrey Brewin, until the spring of 1994 product engineering manager at Electrolux, explains how they now lay great emphasis on teamworking and building relationships. As a result, a new product development project at Electrolux Floorcare will usually involve people from engineering design, industrial design, electronic engineering, customer service, marketing, manufacturing, materials and quality departments. The key task for the project leader is to pull that team together to confirm the objectives that they are working towards and to monitor the progress as the product design and development advances.

Brewin believes that making a product development team work is one of the hardest tasks of management. Often different departments are at loggerheads and there is always a tendency for people to project their own professional views and to sit on their professional status rather than take a detached view. But the team leader cannot allow that to happen: 'in a team everyone has to justify their opinions. For example, a marketing manager cannot behave as if he or she knows all there is to know about marketing and that their comments should be accepted uncritically. They have to convince the team and win the argument.' These team members should not make pronouncements based on their own personal preferences, because most, if not all, of them are likely to be enjoying a lifestyle that is not representative of the general public.

From a purely design point of view Electrolux Floorcare's approach now focuses on the importance of four principles: bringing in industrial designers at the earliest possible moment; using CAD-based visual tools to describe the new product; encouraging creativity; and building sufficient tolerance in a design. The development of the Airstream range of vacuum cleaners demonstrates the importance of building a good relationship with an industrial designer. For this range the company brought in John Besford, formerly from the Fitch consultancy and now at DCA, at the very beginning. Bringing in industrial designers at the start is seen to be critical because almost always there have to be trade-offs between the industrial

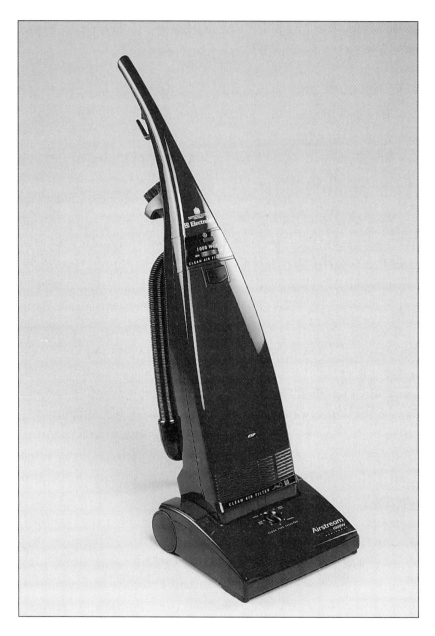

A major part of the success of Electrolux Floorcare's Airstream range of products stems
from building a good relationship between the company and its industrial designers.
(Electrolux Floorcare)

design and the engineering design. This relationship worked well because the consultant designer understood extremely well how the products were going to be manufactured, and took pains to talk to everyone to understand what they were doing. Another advantage was that the engineers and designers at Electrolux were able to prepare very clear product specifications, which meant that Besford's original sketches for the products hardly had to be changed as they neared production. The excellence of the engineering and the design were vindicated by the sales and the whole project was completed fairly rapidly.

The use of advanced CAD systems has produced the biggest change in the design process at Electrolux. They have enabled the company to produce accurate prototypes months before production starts; to supply very detailed information to its toolmakers; to have detailed 3D pictures that both designers and engineers can understand; and to derive pictures of the impending product for the marketing department so that it can plan its marketing campaign well ahead of the product launch. The new technology has changed many relationships, such as those of engineer and toolmaker, engineer and marketing manager, and engineer and industrial designer, with the result that there is interaction between different people very early on in the design process.

CAD systems have been particularly valuable for industrial designers, more at home with three-dimensional pictures than with two-dimensional ones or even drawings, and also for communication purposes because, while engineers feel comfortable with drawings, three-dimensional images communicate much more than drawings to most people. As Brewin explains: 'Putting things into a drawing is like writing an article in Latin and then giving it somebody who has to translate it and put it back into English in their heads. Engineers do that all the time, and they do that in their head, but you can't tell somebody what is in your head.'

Electrolux aims to allow designers and engineers the freedom to be creative, which also means the freedom to make mistakes. But safeguards must then be built into the product development process: while encouraging someone to go out on a limb with a particular design idea, one must ask them to have a fall-back design in case their own pet idea fails.

Another principle which guides the company is ensuring that designs can be manufactured easily, consistently and repeatedly. Their aim is to follow the Taguchi argument that a design should be 'robust', in other words able to withstand the daily vagaries of the manufacturing processes. If the manufacturing processes 'wander a little', the product's design should be able to tolerate these small deviations. This is especially important for products that use plastic injection mouldings, because manufactured products using that technology do not come out exactly the same size every day. To the naked eye they may be the same, but there may well be deviations.

Perhaps the biggest challenge for manufacturing companies is speeding up the process of design and development. In the fast-moving domestic appliance markets fashions change quickly and companies are vying with each other to bring out novel products. Companies such as Electrolux and Black & Decker are working to tighter and tighter time-to-market schedules.

The pressure to bring out new products much more quickly has put an enormous burden on designers and engineers. In these circumstances the time in which everything is done has to be squeezed. The time set aside for deciding on the concept, completing the design, the toolmaking and other activities has to be shrunk. The first essential is to get the market specification right at the very beginning and to establish certainty quickly. Another key point is to stay close to suppliers, ensuring that their manufacturing line is working all the time, that idle time is avoided and nothing holds them up. This is possible with proper integration of both parties' operations.

For a swifter design and development process even meetings have to be more disciplined. Those involved have to prepare for meetings and make decisions at them rather than after them. The need for speed underpins the need for teamwork. It also means that creative ideas have to be encouraged, and then explained, shared and accepted.

Steelcase Strafor

Steelcase Strafor is the world's largest manufacturer of office furniture and office systems. Inspired design and effective marketing have played a large part in its success. In its UK operations this is particularly evident in its Gordon Russell subsidiary.

Gordon Russell was founded in 1927 and quickly became a byword for good design. Gordon Russell himself was steeped in the late nineteenth-century Arts and Crafts movement, which led the graphic and fine arts away from the ornate Victorian style towards a new style that was cleaner and simpler. Gordon Russell's furniture was functional in the best sense of the word, hand-crafted and pleasing to look at. The company remained in the family until 1986 when it was bought by Giroflex, a Welsh office seating firm, which in turn was bought by Steelcase Strafor in 1989. Steelcase Strafor is a joint venture between Steelcase in the USA and Strafor in France. Their principal activities are in metal and laminated office furniture.

Under its new owners, Gordon Russell has expanded into

Steelcase Strafor's emphasis on ergonomic design has been a key to success.
The company researches how project teams use furniture and how people personalize
office space. (Steelcase Strafor)

office systems, providing a complete set of office equipment
which is all made out of wood. The bespoke work on which the
company made its reputation remains, but having also taken the
radical route to produce higher volume products, the company
has had to think hard about what designs will be attractive and
apposite in today's modern office. The result has been a series of
products, that not only look good in themselves but which can
also sit easily alongside the company's other products. In
addition they have been designed to accommodate desk
computers easily and to incorporate layers of cabling. These
desks, chairs, partitions and cabinets have proved popular not
only in the UK but also in Continental Europe.

The key to this success says Dominic Artaud, chief executive of Steelcase Strafor plc, lies firstly in ergonomics. This is where all the group's designs begin: is a product easy to use and is it comfortable to work with? Then the technical aspects are considered such as lighting needs, installation requirements and how a particular product would fit with another one – could a newly designed desk be attached to a partition or another desk and could it accommodate an attachment of some kind?

This accent on ergonomics is further explained by Paul Wade, one of the company's marketing directors. He says: 'About 70 per cent of a company's costs are derived from its people. Therefore it is important to create the right working conditions for them. So we look at the way people use computers at their desk, at the way project teams use furniture and the way that people like to personalize their own office space. All these aspects influence our designs.'

The trend among organizations towards using project teams to solve problems has been especially influential in the design of Gordon Russell's products. The company has designed a number of desks and attachments to enable small groups of people in an office to have the best of both worlds: through the use of low and high partitions as well as differently shaped desks, a combination of the company's products allows quiet work spaces and easily accessible public spaces that can form the focus for a meeting all in a fairly confined area.

A cluster of ideas inform the design of all the company's products, explains Artaud: 'Besides ergonomic considerations, we aim for simplicity, coherence, optimization of space, modularity, flexibility, opportunities for customization and upgradeability.' In order to achieve these objectives the company uses external designers as well as its own and also works with architects. 'The architects play a major role in the development of new products, because they can show what is feasible and what is not. We are not just interested in producing beautiful furniture but in making furniture and systems that will fit easily into a modern environment.'

Although Dominic Artaud says that the company's designs begin with ergonomics, in another sense they begin in the research and development department and in marketing. In the UK, Steelcase Strafor has eight people engaged in research and development, feeding ideas into Gordon Russell and two sister UK companies, while in its European headquarters in Strasbourg it has 30 people in R&D. Marketing information plays a very significant role in the design of the products. Artaud explains: 'We have 200 dealers in the UK with whom we are in constant touch. They can see what is selling, what influences people to buy one product more than another, and they see what our competitors are selling. But we also use market research organizations in order to gauge long-term trends. The total research is methodical and practical, and we put it all together for our designers.'

As well as emphasizing good design, the company has put much effort into improving its production quality and its manufacturing methods. A multiskilling programme for shop-floor workers began in 1992 and enables workers to switch easily from making one kind of product to another. Teams of workers have been given overall responsibility for quality; they have been given full authority not to despatch products if they are faulty.

This training and quality programme paved the way in 1994 for Gordon Russell's two manufacturing centres to be brought under one roof in Broadway, Worcestershire, which meant that bespoke furniture and high-volume products were being produced side by side and by the same workforce. Quality has been ensured by raising the standards against which suppliers are assessed. As Artaud says: 'We are very tough on our suppliers because we want to have perfect products.' Finally, the company has upgraded its manufacturing methods and introduced new concepts such as Just-in-Time manufacture, but some parts of the manufacturing process are still done by hand such as the

Gordon Russell furniture, such as the Perform range, shown here, is designed to be functional and keeps to all the best traditions of hand-crafted production.
(Steelcase Strafor)

finishing. When the products are ultimately despatched they may well be sent out wrapped in a blanket so that the customer does indeed receive a perfect product.

The company's brochure for one of its new ranges, Perform, indicates its interest in corporate identity. The cover of the brochure shows a coloured line drawing of an elderly gentleman, looking remarkably like Sir Gordon Russell himself, feeling the shiny surface of one of the company's modern desks with obvious satisfaction. Near him is a computer screen. The picture reflects continuity of history, quality and modernity.

Conclusion

These four companies provide slightly different stories and messages, but one common theme is that they have all tried to bring different ideas and professional perspectives to bear upon the practice of design. They have also realized that a good design can be wrecked by problems in manufacturing, so they have designed for manufacture in mind and they have sought to improve the production process itself.

7 Design for corporate identity

▶ *Why change?* ▶ *Incremental innovation?* ▶ *Timing the change* ▶ *Orchestrating the change* ▶ *Controlled flexibility* ▶ *Being true to reality* ▶ *Maximizing the benefits* ▶ *Managing the identity* ▶ *Identification and differentiation* ▶ *Assessing your company's needs*

WHATEVER KIND of product companies provide, whether manufactured, service or otherwise, and whatever the sector in which they operate, they must communicate effectively and clearly with their customers and support their products with consistent corporate and brand images. A well-designed product will have difficulty surviving on its own: its benefits must be communicated through equally well-designed literature and packaging, and it must fit into a coherent and clearly defined corporate image. Every organization and every business, no matter how small, has a character or personality and that personality will be expressed verbally and visually. The way that the organization responds to the outside world – to ordinary people, big business or government departments – the way that the organization talks about itself and the way that it 'dresses itself', whether through graphics or office design, will all reflect its nature. These outward manifestations of the organization are all part of what is usually termed corporate identity.

Well-designed packaging and literature enhance well-designed products and strengthen brand identity. The Ideal-Standard's Dualux bathroom fittings packaging, designed by David Hillman and Nancy Williams of Pentagram Design (illustrator, George Hoy) are clear, informative and economic. (Pentagram Design)

Wally Olins, a founder of one of the earliest corporate identities consultancies, Wolff Olins, and originator of much of today's thinking on the subject, describes a corporate identity as projecting three aspects of a company: 'who you are, what you do and how you do it.' (Olins, 1990) In his view corporate identity is not an option that a company chooses to have or not to have, and ignoring it will bring problems: 'Every company, however big or small, has an identity, whether it recognizes it or not, and the real question that needs to be faced is whether the organization seeks to control that identity, or whether it allows the identity to control it.' (Olins, 1994).

Commenting on these words recently, Wally Olins says: 'The way an organization is perceived is governed by four vectors: its products – what it makes; its environment – where it lives; its communication – for example, its advertising and PR; and how it behaves – how the receptionist treats you, how the organization answers letters and so on. All these affect the way an organization is perceived, and in most organizations one of those four vectors will be dominant. If you don't control any of those or if you only control some of them, and if you don't see the relationship between each of them, then you will not control your environment. You will not be in control of your identity and the organization will mean different things to different people at different times. So in that sense your identity will be controlling you and not the other way round.'

The most visible parts of an organization's corporate identity are in its products and services, and its visual images – in other words the way it depicts itself on products, stationery, corporate literature, advertising, vehicles and offices. These visual manifestations become the symbols of what the company is. As such they are very important, very emotive and very powerful but their strength lies not only in their intrinsic visual appeal but also in their representational value. What lies behind them is extremely important. What lies behind an image is what will make it memorable. If the quality of the organiz-ation is excellent we will remember that when we see the image, and conversely if

The first hello: the impact of the reception area and the way the receptionist treats visitors is an influential part of corporate identity and can leave a lasting impression. Reception by Wolff Olins for Orange. (Wolff Olins)

Castrol vehicle livery by Sampson Tyrrell: well-designed and well-implemented corporate identity extends coherently through every aspect of company activity, including distribution and vehicle livery. (Sampson Tyrrell)

our experience of an organization is unfavourable then the image will remind us of that too.

Wally Olins has categorized corporate identity into three types: monolithic – an organization using one name and visual system across all its products (such as Mitsubishi or BP); branded – a company which operates through different brands or companies apparently unrelated (such as Unilever); and endorsed – different companies within a group using different visual systems but identified as part of the group through a visual or written endorsement (Olins, 1990). These definitions highlight the complex nature of a corporate identity and the potential to confuse if it is not designed and implemented correctly.

The value of having a clear corporate identity has increased in recent years for several reasons. The first is that as companies become more diversified their image often becomes more diffuse. Mary Lewis of graphics consultancy Lewis Moberly explains: 'Companies believe they have one image when in fact they have several. This can cause confusion in the minds of their target audience.' The second reason is that as technologies become cheaper and more widely available to manufacturers so the differences between products grows narrower. Differentiation through corporate identity has become more important. In the words of John Grey at identity consultants Halpin Grey Vermeir: 'If you can't differentiate yourself using the harder aspects of your business, you use the softer things – the way you look, the way you communicate and the way you behave.' Wally Olins comments: 'In many industries the products are not that dissimilar. So how do you choose between them? You will probably make your choice based on emotional reasons and those reasons will be governed by the way that the company presents itself. If you as an organization can distinguish yourself and your personality in a way that your competitors can't and in a way that they can't copy, then you have gained an advantage.'

The way organizations present themselves determines how people react to them: design provides the means to communicate the unique characteristics of a corporate personality. The Geffrye Museum's unique keyhole identity, symbolizing the domestic interior and designed by Lewis Moberly, is inspired by the idea of unlocking the past and is shorthand for the personal, friendly and accessible experience which the museum offers to visitors. (Lewis Moberly)

The need to build up what is sometimes called 'brand equity' is another reason given for the increasing importance of corporate identity. This means building up good will or favourable 'capital' among the public in relation to a specific brand or, more often, to the company that provides the brand. Chris Lewis of consultancy BBLM explains how this can work: 'If you as an organization present yourself as a series of positive values and people get to know you and feel comfortable with you, then when you make a mess of something (through poor products or bad press) people are more likely to explain it away as an aberration. Your customers will still buy your products or services because they believe in the company. You may suffer some setback, but it won't be anything as bad as it would have been if they felt neutral about you and felt no loyalty towards you.'

These competitive benefits and value factors which form the basis of how a company is perceived explain why an identity cannot be contained only in a badge, logo or letterhead. It is not a case of simply pulling a new marque off the shelf: an identity must have a coherence across everything a company does. A letterhead cannot be redesigned in isolation from the whole range of company literature, logo and other elements, no matter how urgent immediate practical needs might be as a result of, for example, change of name or ownership.

Why change?

A change of ownership is just one reason for wanting to change a corporate identity. Other reasons include: a merger with another company; a change in the direction or emphasis of a business; a desire to look as up to date as your competitors; a need to regain lost market share; highlighting a better product, better service or better quality; the confirmation and

InterCity's new identity, designed by Newell & Sorrell, had to define the organization's future and reflect the aspirations of the people in it. (Newell & Sorrell)

reinforcement of an internal cultural change; or a need to refocus what has become a muddled corporate image. One of the benefits of recreating an identity is that it enables the organization to re-examine what it is in business for. Hilary Boys of Lewis Moberly finds that the process which a company goes through to reach a new identity is as important as the new identity itself: 'It is often a spur for bringing the organization together and thinking through how it should approach its market.'

The introduction of a new identity can also revitalize a company, refocus and motivate staff, and help to encourage changes in cultural attitudes. This is exemplified by the corporate identity for British Rail's InterCity division. In 1986 the division was losing £125 million a year. Morale was low and the organization had just been told by the Government that it had to be in profit by 1988/89. InterCity's director, Dr John Prideaux, decided to build a strategy based on quality and growth and at the same time to commission the consultancy Newell & Sorrell to generate a new corporate identity which could act as a stimulus for the change in corporate culture. The change in InterCity's corporate identity was a substantial one. Every aspect of the business was subjected to a review and improved. John Sorrell explains: 'The identity change was based on making a large number of improvements to such things as the interior designs, the catering, the on-board service, the food packaging, the clothing and the way the staff were trained.' One innovation was to have liveried staff on the platform directing people to the train. The identity change worked extremely well – InterCity was in profit by 1989 – because the visual identity change went 'hand in hand with a much better service.'

This example also shows how the right identity change defines an organization's future as much as its present. InterCity's new identity could not reflect the present. It had to reflect the aspirations of the people in the organization, to suggest where the organization wanted to go in the future. It became a rallying point for the employees and made the idea of change and the objective of a much smarter, more efficient service accessible to those inside the organization.

Incremental innovation?

There are three requirements for getting the most out of a new corporate identity: involving people, using the visual identity imaginatively and avoiding radical change for the sake of it. A revolutionary shift in corporate identity is not necessarily the best form of change: some of the most effective changes are the most subtle. Brian Webb of graphic design consultancy Trickett &

New or revitalized identities can help to refocus and motivate a company and its staff. Rapid growth at Eagle Star had led to a fragmentation of its identity and many consumers were unaware of newer financial products. Lewis Moberly redesigned the visual identity to convey power, strength and stability. (Lewis Moberly)

Webb comments that with some of the more traditional visual identities such as those used by ICI, BP or Shell, major changes are unnecessary. All three companies made some minor changes to their logos in the 1980s which were successful. To some people the change that these companies made to their logos seemed cosmetic and expensive, but that is to miss the point. These small changes were useful for two reasons. Firstly, they signalled that the companies were not stuck in the past, that they were prepared to adapt, and secondly, the process of change – the consensus-building – was very valuable.

Even with less established names, the argument for a minimalist approach can be valid, says Webb. 'Sometimes we say to a client, you don't need a radical change. All you really need is a wash and brush-up, that is, you need to keep the good bits of the identity that you have built up and improve the others.' This is especially true of fast-growing companies, where the corporate identity is lagging behind the rest of the business so that it reflects the past much more than the present, let alone the future. 'What you are trying to do with your corporate identity is to match the quality and image of the company with the quality of the product. So it may just be a question of saying in visual terms that your quality is even better than it was.'

Timing the change

The timing of an identity change is clearly very important. According to Olins a good time is 'when something important is happening to you, for example when you are launching a range of completely new products, or when the products in your market are homogenizing, or when your visual styles are getting out of date.' He also argues that a time of great stress can be an appropriate time in which to reassess and recreate the identity, provided that the company is clear about its ultimate goal. It is arguable, however, whether an identity change should follow or coincide with a corporate change. Olins believes that thinking about your corporate identity while you are also reviewing other major issues such as strategy works well: 'A corporate identity review can have a catalytic effect. You have to think about what kind of organization you are and are going to be when you examine the corporate identity, whether you wish to be a single focus monolithic organization or whether you will strengthen the subsidiaries or accentuate the separate brands and so on. In addressing those issues you will be led to think about others. So a corporate identity change can have a galvanic effect on the organization.'

In Olins' view an inappropriate time to introduce a new identity is when there is no real reason to do so, for example when the company's position in the market is assured, or if a product or service is so bad and struggling to maintain its position that a corporate identity change would only highlight deficiencies. Others argue against an identity change when employees are so jittery that any move to bring in outside consultants would be counter-productive.

Orchestrating the change

Examining the aims of a business is central to reshaping its identity. Companies must also have a clear understanding of why they want to reshape it. John Sorrell explains: 'It is incredibly important that you establish with the organization what its objectives are in making the change. It needs to be clear what it is trying to achieve. But it is not just a question about where it is going now, but what its aims are for the long term. It is absolutely critical that the organization projects itself into the future, so that you can shape the corporate identity in such a way that it can move forward and be adapted later.'

Too many companies still take a short-term view of corporate identity and look for quick solutions to one element of it rather than assessing the long-term value of the identity as a whole. Many of the corporate identity changes of the 1980s are seen as somewhat superficial for that reason: they reflected the fashionable design styles of the 1980s rather than communicating a deeper message about the organization.

As well as thinking about the long term, an organization bent on changing its identity must research how it is perceived in the market. Consultants often include this as part of their audit of a company's strengths and weaknesses. When Sampson Tyrrell created a new identity for two merged companies in the Burmah Castrol group – Fosroc and Expandite – it conducted an extensive audit among the customers of the two companies.

The audit showed that Fosroc had a much more go-ahead image than Expandite. As a result the Burmah board agreed to Sampson Tyrrell's suggestion that the merged companies simply be called Fosroc.

The loss of a subsidiary name within a much larger organization is almost always a difficult decision and can often stir up political conflicts which may be difficult to resolve. It is therefore essential to involve and win the support of every level of the organization when establishing a new identity, but employee involvement has its limits and there comes a time when a decision has to be taken to adopt a particular corporate image. The tricky part of this process is balancing the power of the senior managers against the power of divisional or brand managers and other employees.

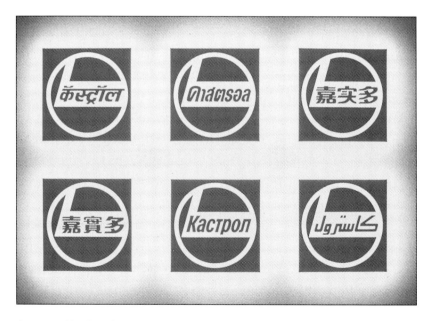

Corporate identities for multinationals demand both international coherence and sensitivity to local conditions. The Castrol logo, while reflecting the culture of a particular country, is undeniably still a Castrol logo. (Sampson Tyrrell)

Controlled flexibility

A poorly executed identity change can lead to chaos in the long term, with subsidiaries or key individuals refusing to support the new identity or undermining it. Mary Lewis of Lewis Moberly cites one company where within two years 'virtual anarchy' resulted from a major change of direction and new corporate identity. Later, when Lewis Moberly were called in to help, it was clear that various parts of the organization were issuing their own style of literature yet still using the corporate logo. The managers justified this on the grounds that they were serving discrete market segments. In reality these market segments overlapped, so the messages coming out of the organization were confused. The situation was out of control and the design department lacked the status to impose its will. Lewis Moberly advised the company to counsel staff, to share their concerns with their employees and to show them how they could still issue their own material in a much more coherent way.

David Allen, managing director at Sampson Tyrrell, advises companies to implement a new corporate identity with great flexibility, especially if it is operating across several countries: 'With multinationals there is a need for both more international coherence and more reactivity to local conditions and that creates enormous tensions for identity management. So you can't take the top down approach. That doesn't work in many organizations any more where you have divisional directors with their own profit centres, who know more about their areas than anyone in head office and who will want to use the company's visual identity in the way that they think best. You have to provide a different way of giving a sprawling company visual coherence. You have to decide what things should be common to every part of the organization and what things you can be flexible on.'

Wolff Olins's work for Forte allows the company's various hotel services different visual identities under the Forte umbrella. (Wolff Olins)

The way in which some companies have attempted to lay down rigid corporate identity rules runs flat in the face of other initiatives to make people in those organizations more responsible. If 'empowerment' and 'delegated responsibility' have any meaning, then it must include the freedom to use the company's visual identities in a flexible way. Retaining individuality within one company is not easy, but there are, as John Sorrell explains, a host of subtle ways of 'endorsing the relationship between subsidiaries and the parent company.' The subtlety lies in making small shifts in the design for a particular part of the group. Sorrell also emphasizes that 'the corporate fist' approach can often not only alienate those inside an organization but also those outside it, while customers also appreciate individuality.

Corporate identity consultants and companies themselves

must, however, share responsibility for the way in which companies' visual identities have often been steamrollered into place without room for flexibility. As Olins himself reflects: 'In a way the consultants have been hoist by their own petard. We've argued for homogeneity and our clients have gone down this route. But now we've moved on. We're all saying that you can't have rigidity because you need to respect individual cultures. You can't be the same everywhere because that is alienating and depressing. What we are now arguing for is some differentiation within the corporate identity, but it has to be in an ordered framework.' In John Grey's view too many people believe that you have to apply the visual identity in a totally uniform way. This becomes boring and restrictive and absolves people from looking for new ways to portray the corporate identity message. The identity, like the business itself, must develop and innovate.

One of the worst ways of implementing an identity change is simply to establish a book of rules which dictate how the new logo and other visual appurtenances should be used. Mary Lewis points out that this can be dangerous because in effect it means that the company has stopped thinking. It can also mean that the company focuses its attention on the most obvious visual elements and neglects to think how it is extending the identity handwriting. More than that, a rigid rule book stifles personal interaction with the identity on the part of the users.

Being true to reality

A corporate identity has to tell the truth to be effective. The visual image itself is meant to influence how we see the organization and to that extent to shape our experience of it, but there are limits as to how far the visual devices and messages can make us feel good about a company when in fact we think the opposite. The identity 'has to come from within' and be both based on and inspired by what the company feels about itself. If this is not the case the new corporate identity will fail. Even though most corporate identity changes – at least the visual components – are based on the work of consultants, the inspiration for the identity and the messages that it will enshrine should still come from the company. Brian Webb defines vividly how this should work: 'A good design should be like a suit: if the client company is not comfortable wearing it, if it doesn't fit, then it is not right.'

Consultants and their clients must understand the possible consequences of a corporate identity and be sure that the new image truly reflects what is behind it. John Sorrell emphasizes the importance of this: 'An organization that wants to use corporate identity in a significant way has to look at the way that the identity will tell the truth about the organization. You can't lie with corporate identity. An organization that is hopeless cannot paper over its incompetence with a new symbol and think that it has changed its identity. People will find them out.' What such an organization has to do is first change itself.

Although a new logo and colour scheme are dominant elements of a corporate identity they are in fact the tip of the iceberg. It is not the visual changes but the changes to products, services, behaviour or culture that are really significant. When British Airways changed its identity in the 1980s it worked because they upgraded the service before making outward visual changes and embarking on a new advertising campaign: they supported their claims with genuine improvements. Nor is a visual identity completely encapsulated in the marque. There is, as Hilary Boys of Lewis Moberly explains, a visual handwriting attached to a corporate identity, which most companies fail to

New identities can communicate qualities that might otherwise remain invisible. When Computer Cab invented a credit card system Trickett & Webb designed the credit card, the annual report and the promotional material. One of the reasons for the new identity was to reflect the company's advanced, efficient operation. (Trickett & Webb)

Literature provides a means of communicating a company's image, objectives and activities powerfully. Shown here: brochures for Gestetner, Buchanan and Akai, designed by BBLM. The Akai proposition, 'Serious Pleasure', is a key element of the company's identity and promotion (see page 148). (BBLM)

see: 'Marks and Spencer is a business with a very definite handwriting which pervades everything it does. It has certain ways of displaying its products with various kinds of promotional material which if done differently you would immediately notice. An obvious example would be if they advertised their ladies' underwear with a model posed in a provocative or suggestive way. If on the other hand you look at a company that does not think about its identity so thoroughly, you will see things in its brochures or advertising that don't quite fit in with their corporate identity. That is because such companies think that all they have to do is to follow the guidelines about how to use the logo. So there is an organic approach to using an identity which most companies don't appreciate.'

Maximizing the benefits

The success of a corporate identity will also be undermined by an inept launch and half-hearted or mediocre implementation. One problem that companies have experienced is that they launch a radical visual identity change without preparing their employees or their customers sufficiently. This causes negative reactions to the new identity: the change is too sudden and therefore unwelcome.

Mediocre implementation can also be very destructive. Many commentators – in this field as in others – say that while UK companies are eager for new ideas they are not good at putting them into practice. The problem is that the new idea is often seen as an easy panacea, what Mary Lewis calls 'a convenience tool' – something that one can buy ready made, plug in and walk away from. But implementing a new corporate identity takes considerable time and it has to be done carefully and thoroughly. Hilary Boys finds that many companies underestimate the time and the cost of implementation: 'With a lot of companies they stop at the presentation stage or when they have produced their corporate identity manual. The result is that two years later, they are likely to ask themselves why the change was not as effective as they had expected'.

The potential of a visual identity is often missed. It can be a major means of promoting a company in the wider world as well as in the business community or among its customers. But the context of the promotion must fit in with the corporate identity. The company's products should sit happily in the promotional environment and not look out of place in the way that cigarette advertising looks odd at a sporting event. BBLM's Chris Lewis recalls creating a promotion for the Japanese audio electronics company Akai. This centred on a competition with prizes including a chance to learn to water-ski and the opportunity to go ballooning. These prizes gelled with the company's corporate identity, says Lewis, because they mirrored the company's proposition – selling 'serious pleasure.' (See page 146.)

Managing the identity

A corporate identity is not something that should be thought about every five or ten years and then forgotten. Ideally it should be part of managers' consciousness; it ought to evolve and this means that it needs to be managed. John Sorrell is one of those who maintains that it needs managing on a daily basis: 'Corporate identity is something that an organization should look at all the time. It should be managed from a senior level, so that every day the identity is managed successfully and appropriately and is moved on. The reason that an organization should do this is because every day it is making decisions that affect the way that it is perceived. These decisions include:

rearranging the reception area, buying new equipment, designing new products and decorating the offices. They are all connected with identity. In fact everything that the organization does has some relevance to its identity. So the corporate identity ought to be managed and focused around one person rather than as present in most companies – managed by lots of different people. If you look at a company like British Airways, they do manage their identity every day.' If the identity is managed properly then it will enhance the reputation of the organization and enhance the culture of the organization.

Wally Olins maintains that the corporate identity should be managed by someone who has a power base in the organization and who has funds to achieve his or her aims. Yet the difficulty, as he admits, is that a company's identity is enshrined in four facets of an organization: its strategy, its marketing, its communication and its behaviour. One person or maybe one department ought to manage the corporate identity from these four angles to ensure that it is always developing and kept fresh.

Identification and differentiation

Some firms have adopted a very unimaginative approach to implementing the designer's original ideas. This can mean that one bank looks much like another bank and one international accountancy firm also looks much like another. Yet the point of design is usually to differentiate not to follow. One reason why there is sometimes a similarity between the visual identity of one firm and another is that some firms want this, particularly in the case of packaging design. Brian Webb recounts that clients will sometimes come armed with examples of what their competitor is doing and say, 'I want something like that.' This is particularly true in the retail sector.

Whether on letters, on buildings, on advertisements, products or vehicles, logos often create the very first impression. Shown here: old Prudential identity, and new design by Wolff Olins; old BT identity, and new design by Wolff Olins; red and green logo designed by BBLM for Premier Marque vintage car restoration company (for whom BBLM also devised name branding); logo designed for Mainline by Halpin Grey Vermeir when the railway freight company was privatized; NPI identity design and implementation by Coley Porter Bell.

Designers fight against the 'me-too brief' as much as they can, arguing that a client should want to stand out from the competition, but it is often an uphill battle. Mary Lewis comments: 'Many companies pitch their sights rather low. They are wary of being adventurous. We work with our clients encouraging them to keep an open mind, to be prepared to learn as we move through the process.'

Wally Olins explains that most industries tend to have a generic visual style, which the individual players in that industry find very difficult to break away from. In his book *Corporate Identity* he says: 'A generic is usually created by chance, by what the leader in the industry happened to choose to do at a certain point in its development because it seemed right for the circumstances.' Then peer group pressure takes over and everybody wants to look like the leader whether it is appropriate or not – and in most cases it is not. However, being seen to be very different takes courage, as Olins explains today: 'When you suggest a radical departure in visual identity, you are asking the person who has responsibility for agreeing to it to take a decision which has profound implications for the way people will see the organization and for the way people will see them as chairman or chief executive. That is risky and it requires courage to take that risk. If you take a radical step you are going to get criticized.'

The unusual visual identities that Wolff Olins created for BT and the group renamed Investors in Industry (now known as 3i) were deliberately very innovative. BT's aimed to convey that the organization had embarked on a new path; the 3i identity communicated that the organization was highly distinctive and imaginative. Yet both identity changes, especially BT's, attracted much criticism. Olins justifies the changes: 'We were trying to show that BT was changing from being a bureaucratic, UK-oriented organization to one that was becoming international and communications-based. If you look at BT today it is unrecognizable from what it was in the 1980s. It has improved immeasurably and it is operating as a global company. That dramatic change needed to be encapsulated in a totally new visual identity. It was inevitable that the company would be criticized, but it had to be done.'

Assessing your company's needs

Any organization thinking of changing its corporate identity needs to make a number of probing assessments beforehand. Is the identity renewal really necessary and is it appropriate? If your company is experiencing an identity crisis, perhaps it can be solved just by talking to your employees, rather than papering over the cracks with a new logo. Make sure that you understand your company's long-term objectives and why you need a change of image, and take responsibility for what you are doing rather than leaving everything to a consultant without thinking through all the consequences. And finally, in David Allen's words: 'Recognize that your identity is of value, and that your identity is to a large extent what people will judge you by.'

Planning a new corporate identity

- Is it just a question of visual identity or is it something more: do you need a design consultant or a management consultant?
- Evaluate and understand internal and external perceptions of your business and its identity.
- Set clear and measurable communications objectives.
- When choosing a design consultant initially, consult design registers and industry associations. Develop a shortlist of three possibilities, agree selection criteria and be prepared to pay for written and creative proposals.
- Divide the identity change process into stages:
  - the initial research of current perceptions and the size of the task;
  - the creative development of the new identity and its basic elements;
  - the implementation of the new identity through all items from stationery to signage and the production of guidelines.
- One firm could handle all three stages but it may be worth splitting the programme: for example, you could commission independent research.
- Research and strategy development are important but make sure you evaluate the competing consultancies' creative strengths.
- Plan the introduction of the new identity and how it will be managed.

(Source: John Grey, Halpin Grey Vermeir)

8 Conclusion

▶ *The role of individual businesses* ▶ *Education*

DESIGN IS OFTEN the key factor that makes a product or service successful. If a product is attractive, appropriate to its market, works well and is also economic to make it has won more than half the battle. In the service sector success depends to a considerable extent on how the customer services are designed and communicated; on corporate literature – whether it is informative, attractive and welcoming; and on the whole corporate image – does it convey an accurate and inviting picture of the company? Yet the power and influence of design is still frequently underestimated.

Perhaps the biggest reason for this is that in many companies the existence of design is elusive. It is everywhere and nowhere at the same time. Design is likely to feature across a range of activities often unconsciously with many people engaged in design decisions without knowing it. When marketing managers propose a change in a product or service they are making a design decision and when office managers buy a new reception desk that too is a design decision because the purchase will affect the corporate image.

Many people, be they in marketing, customer service or public relations, think they can design. They may have some good ideas on design, but their training is unlikely to have included design to any significant extent. In other words they are not experts in design. This is not to say that every decision with a design content should be passed to a designer, but it does mean that we ought to be aware of the role that design plays in daily business decisions so that in important instances the decision is influenced by someone who has some design training. At the same time these key decisions (which have a design component) ought to evolve within the context of the company's general business philosophy.

One characteristic of the companies that featured in the case studies was that they had given design a recognizable place in their organizational structure. In the case of Rover, product design is the hub of everything that the company does. The same is true of Bluebird Toys. These companies are using design as a focal point or in the words of the big US electronics corporation, Hewlett Packard, as the glue in the business. They think about how best to design a product for a customer, how best to design it for manufacture and how best to package it. Moreover, these decisions are taken together so that they support each other and the totality of the decisions taken are made within a clear financial framework. No-one in these companies talks about design as just a pretty face without any regard to product performance or profits or customer preferences. They are making their decisions on clear practical grounds.

At the same time these companies do not give designers pride of place at the expense of other professionals. What they are keen to do is to make sure that the voice of 'the designer', is heard on an equal footing with that of others. One of the weaknesses of British management has been to give an unnecessarily high primacy to one particular profession. In some companies it has been engineers, in others it has been marketing people while in many companies it has been and still is accountants. All the companies featured in this book stress the importance of teams. These teams do not operate as collections of prima donnas from various management departments where territorial positions are stated and lengthy and difficult compromises follow. The people in these teams try – and they appear to succeed – to look at major management decisions from an overall business view, in which a design perspective plays a big part.

These companies also spend a considerable amount of time ensuring that the process of decision-making and the implementation of decisions is working. As Chris Floyd at the international consultancy Arthur D Little so rightly pointed out, the reason why teamworking often founders is that it is not given the support that it needs from above, and therefore not given the support from the rest of the organization at critical times. The success of a team can be judged by how well it works at times of urgency not just by how well it works on a day-to-day basis. This is especially true of those companies where some managers wear more than one hat.

In the 1980s a number of companies – especially in the service sector – saw design as a relatively cheap and cosmetic way of winning customers. But design that is merely used to cover up what is underneath is of no value in the long term. The quest for corporate re-imaging has been intense, but in many cases it has failed because what lies behind the image has not improved. Also, in some cases the new image has been too radical and has jarred with what customers have been used to. However, some companies such as Boots, Marks & Spencer, W H Smith and British Airways have used product design and corporate identity design in a coherent way by making design decisions within a clear, customer-led business strategy.

If more companies are to use design imaginatively in their

businesses, change must come from three sources: the business community, the business and design education practitioners, and from design organizations such as the Design Council. In addition, these three groups have to work together with a common purpose (which has not been evident so far) to get the message of good design across. In this context good design means good design for a business result not good design for the sake of design.

The role of individual businesses

For companies that have not yet seriously embraced design ideas, the messages in this book might seem daunting as well as expensive. Where do they start? How can they find out what they need? How can they introduce some design concepts into their business without spending an inordinate amount of money? Wendy Powell, Head of Design Management at de Montfort University and an adviser on education and training policy to the Design Council advises companies to test the benefits of design in stages. Her first suggestion is simple, practical and short-term: send staff on two- or three-day courses on design awareness. A short exposure to design ideas and practices can have a strong effect on participants. It can help key figures in an organization to understand design and then explain its significance to others so that eventually everyone in the organization appreciates design. Just bringing in an outside designer without preparing the organization for this, which is what some companies do, can work very badly. The 'not-invented-here' syndrome often means that designer-inspired ideas are rejected. People in companies have to see for themselves why design works and feel comfortable enough with designers' way of thinking to be able to question it when necessary.

Wendy Powell also suggests that a designer could be planted at one or two key points in the organization, for example perhaps at board level as a non-executive director. Although he or she need not necessarily have to give advice on design-related matters, the designer could give a designer's point of view on a range of non-design issues, showing how valuable design input can be. If that worked, a designer could be introduced into the organization in a more formal way.

Companies can also introduce design concepts into their organization by hiring designers on a part-time basis, but on two conditions. Firstly, the external designer has to be well briefed about the organization, its senior people, its history and its culture, and the designer must be prepared to adapt his or her methods to suit the organization. Secondly, all those people with whom the designer will work have to be consulted and prepared. Bringing in an outside designer should not cause any people in the organization to feel undermined or threatened. The external designer's skills should be seen as complementary to those within the company rather than being superior or contradictory to them. This in turn means choosing a designer carefully. As Sir Terence Conran emphasizes in Chapter 4, choose someone who speaks your language, who thinks as you do. That in turn means being sure what you need. When Peter Gibson, chief executive of the Surgical Technology Group, hired designers on a temporary basis, this worked partly because he and his engineers knew exactly what they required in the context of a project with a clear aim.

One reason why companies can be reluctant to spend money on design as a formal activity is that they worry about the reaction of their financial backers, whether they are banks or City institutions. This could be remedied in two ways. Firstly, when good design works for companies, they should make this clear in the annual report. Very few companies ever mention design in their annual review of their business. They talk of markets, new products, sales, overseas competition and profits, but the impact of investment in design is rarely mentioned. Too many company reports are unrevealing in that while they may refer to product launches, they give little clue to the effort and expenditure that lay behind them. It is not surprising that backers underestimate the importance of investment in innovation and design.

Secondly, companies ought to counter head on the charge that investment in design is unquantifiable. Financial experts routinely make assessments about other aspects of companies, such as marketing capability, product performance and top team competence, but, as Wendy Powell points out, 'If they can evaluate the quality of the management team, why can't they evaluate the quality of the design? Accountants say to the Design Council that they are not trained to evaluate design, but I always reply, "how do you assess a company's marketing ability, how do you assess a company's products – are you trained to do that?"'

Education

As is so often the case, the root of the answer lies in better education at professional level. Design hardly figures in business education, and business perspectives must feature more prominently in design education. The narrow sectionalism of UK management and professional education is highly damaging. For example, it really seems nonsensical to study marketing without design, since the two subjects are so closely interwoven. Similarly, those trained in accountancy need to be exposed as part of their training to design, marketing and product and service development practices. Otherwise the numbers that they read on balance sheets will remain just that.

Chief executives must educate and train their staff to understand the value of design input and effective design management, but the teaching of design itself must continue to adapt to the needs of business. Design courses are gradually changing to include management and professional practice elements, with the aim of redressing the imbalance in designers' education by introducing business studies to designers, but there is a growing number of people in the design world who feel that much current design education is inadequate. While it encourages excellent creative skills in potential designers it does so far removed from the business environment in which they will have to earn their living.

David Walker of the Open University, who recently helped to found a design management course at Brunel University, points to the prevalence of the 'blue-sky design project'. He feels that this way of learning is very damaging, because it means that when students 'discover that they have to work within constraints, that they have to work with people in marketing and finance and they have to think about customers, they have absolutely no idea how to deal with this set of factors.' If heads of design colleges emphasize design in isolation rather than in the context of business success, the design message is undermined.

Designers who have succeeded in the business world have often had to learn on the job, the hard way, how their own design expertise has to be tempered by differing management and market needs, but not everybody has this opportunity. Many leading designers believe that design education should be rethought, but management and business training should adapt in parallel. Although much has already been achieved, stronger links between the business world and the design colleges could help to remedy both sides of the problem.

John Sorrell recognizes that change is necessary and believes that there is an unprecedented 'climate for collaboration' amongst the many design-related organizations. One of Sorrell's aims is to foster a spirit of dialogue between the numerous organizations in the UK that have an interest in design and whose accumulated experience could be galvanized to promote the importance of design, and develop design education to fit the needs of the next century. If design is linked in everyone's mind with technology, science and innovation, rather than being something peripheral, the battle will have been won.

Above all, what is needed is for the business world and educational establishments to catch up with public opinion and public sensibilities. More and more people do appreciate good design. People buy a great many products and services on the strength of their design. Many managers instinctively know this to be true when they are outside their business environment, yet once they walk through the swing doors of their offices this message seems to get forgotten. But design is fundamental to business survival; it cannot be overlooked.

References and bibliography

Arthur D Little, *Management Perspectives on Innovation*, Cambridge, Massachusetts, USA, 1985

Berliner, C, and Brimson, J A (eds), *Cost Management for Today's Advanced Manufacturing: The CAM-I Conceptual Design*, Boston, Harvard Business School Press, 1988

Birchall, D, and Swords, S, *Growth and Innovation – The Golden Triangle Business Survey*, Henley-on-Thames, Henley Management College/Price Waterhouse, 1993

Connell, D, 'The UK's performance in export markets – some evidence from international trade data', *Discussion Paper 6*, London, National Economic Development Office, 1979

Constable, G, *Managing product development – six case studies (Practical application of BS 7000)*, published by the Design Council for the DTI, London, 1991

Fairhead, J, *Design for a corporate culture: a report prepared for the National Economic Development Council*, London, National Economic Development Office, 1987

Farish, M, *Strategies for World-Class Products*, Aldershot, Design Council/Gower Publishing, 1995

Kravis, I and Lipsey, R E, *Price Competitiveness in World Trade*, New York, Columbia University Press, 1971

Lorenz, C, *The Design Dimension*, Oxford, Blackwell, 1990

Myerson, J, *Building Bridges – A Study of UK Postgraduate Courses in Industrial Design Engineering*, London, Design Council, 1993

Oakley, M (ed), *Design Management: A Handbook of Issues and Methods*, Oxford, Basil Blackwell, 1990

Olins, W, *The Wolf Olins Guide to Corporate Identity*, London, The Design Council, 1990

Olins, W, *Corporate Identity – Making Business Strategy Visible through Design*, London, Thames & Hudson, 1994

Open University/UMIST Design Innovation Group, *The Benefits and Costs of Investment in Design*, Milton Keynes, Open University, 1990

Pilditch, J, *Winning Ways*, London, Harper & Row Ltd, 1987

Pugh, S, *Total Design – Integrated Methods for Successful Product Engineering*, Wokingham, Addison-Wesley, 1991

Roy, Dr R et al, *Design and the Economy*, 2nd edition, London, The Design Council, 1990

Schott, K and Pick, K, 'The Effect of Price and Non-Price Factors on UK Export Performance and Import Penetration', *University College London Discussion Paper No 35*, London, 1983

Smith, P G and Reinertsen, D G, *Developing Products in Half the Time*, New York, Van Nostrand Reinhold, 1991

Ughanwa, Dr O D, and Baker, Professor M, *The Role of Design in International Competitiveness*, London, Routledge, 1989

Index

Design Protection
Fourth Edition
Incorporating the 1994 Trade Marks Act

Dan Johnston

Consultants: Michael Edwardes-Evans & Geoffrey Adams

A Design Council Title

The importance of design in modern industry and commerce is becoming ever more clearly realised worldwide. And with that realisation comes recognition of the damage done by copying. To combat plagiarism it is essential that designers, managers and company directors understand the range of legal protection available. Patents, trade marks, design registration, copyright and design right are the chief means by which intellectual property rights may be protected. Added to that is the common law right of 'passing off'. All these methods are covered comprehensively in this new edition of the standard work on this subject.

Written by the leading UK expert, this book offers the most up-to-date guidance on design protection, from patenting inventions to protecting styles and brands. It covers the content of the 1994 Trade Marks Act. It also explains design protection terminology and its underlying principles as well as outlining the latest European legislation, likely implications of the impact of EU initiatives in the future, and international differences.

1995   352 pages   0 566 07553 9

Gower

Comparative Contract Law
England, France, Germany

P D V Marsh

Despite the media emphasis on the 'Single European Market', people who do business across the EC are faced with radical differences between legal systems and philosophies. It is dangerous to make assumptions about another country's law.

Peter Marsh's book reviews and compares the main elements of English, French and German law as they relate to business contracts; especially those relating to the sale of goods and to construction work. He covers

- drawing up contracts
- their validity
- the obligations of the parties
- the position of third parties
- the control of unfair terms
- remedies for non-performance.

As the only single-volume detailed comparative treatment of both French and German contract law in the English language this book will be invaluable to British businesses trading with France and Germany, to lawyers who may be called upon to advise such businesses, and to professionals in the construction industry who may be carrying out work in France or Germany.

1994   392 pages   0 566 09006 6

Gower

Copyright Theft

John Gurnsey

Copyright was developed to protect the printed word. In the late twentieth century can it and does it realistically serve to protect authors of audio, video and electronic products that are the vehicles of information supply in our multi-media age?

Systematic copyright theft forms part of a multi-billion dollar international industry, which is able to thrive partly because it is easy to overlook what is known to be theft when the original material remains intact. But what was a 'cottage industry' 30 years ago has now become much more sophisticated, so that pirate books printed in Taiwan flood the markets of West Africa, and audio tapes printed in the Far East appear in Saudi Arabia, Australasia and even Europe. The threat to publishers is alarming, and increasing. The burgeoning of the electronic information industry today makes copyright theft an urgent issue.

John Gurnsey has reviewed all forms of copyright theft, from commercial to domestic, gathering the experiences of a wide range of organizations across book and electronic publishing. Book, electronic, database, audio, video, games and multimedia publishing are all considered along with the question of whether existing laws can effectively serve such a rapidly changing industry.

Copyright law is an extremely complex area: this book is about the abuse of it, rather than the law itself. In helping publishing companies understand more about copyright theft, it might help them to avoid it in at least some of its forms.

1995   208 pages   0 566 07631 4

Gower

Best Practice Benchmarking

Sylvia Codling

Benchmarking is potentially the most powerful weapon in the corporate armoury. It's the technique that enabled Cummins Engine Company to slash delivery time from eight months to eight weeks, Lucas to reduce the number of shopfloor grades at one of its sites from seventeen to four and British Rail to cut cleaning time for a 660-seat train to just eight minutes. In other companies order processing time has been brought down from weeks to days, engineering drawings output doubled and inventory cut by two-thirds.

And yet, in spite of the articles, the seminars and the conferences, managers continue to ask "What is benchmarking?" and "How do we do it?" The purpose of this book is to answer those questions. Through a series of case histories and references it shares the experience and knowledge acquired by benchmarking companies across a wide range of industries. Above all, it provides a detailed step-by-step guide to the entire process, including a complete set of planning worksheets.

Case studies include: Siemens Plessey, Volkswagen, British Rail, Lucas Industries, Shell, Rover and Hewlett Packard.

Benchmarking is a flexible discipline that has become a way of life in some of the world's most successful organizations. Learning from the best can help your own company to become a world leader in those areas that are critical to its performance. In so doing you will achieve an enduring competitive edge.

1995   168 pages   0 566 07591 1

Gower

Dictionary of Marketing

Wolfgang J Koschnick

The globalisation of marketing and the increasing sophistication of the advertising and marketing techniques that are now used have resulted in a massive growth in the jargon, technical terms and specialized vocabulary used in professional marketing. In addition, developments in computers at the PC and mainframe level now make possible the gathering and analysis of data which measures everything from market share to customer satisfaction. All these developments require at least a working knowledge of a wide spectrum of subjects that range from statistical techniques to the social aspects of target markets. The *Dictionary of Marketing* has been compiled to supply the marketing professional with definitions and explanations of the terms used in these burgeoning fields.

The *Dictionary of Marketing* is designed to be the leading marketing dictionary of the English-speaking world. On some 800 printed pages it lists more than 5,000 terms culled from all areas of marketing. It is the most comprehensive reference book available for marketers and students of marketing and related fields. The book is an attempt to set down an exhaustive range of marketing and marketing-related terms and to provide a definition or, where appropriate, a detailed explanation and description for all entries. In cases where a term or a definition was originated by, or is otherwise closely linked with, a specific person, the name is given in parentheses. A large number of illustrations, charts and tables are included to make some of the more complex entries easier to understand.

In selecting the terms contained in this book, the author has defined marketing widely in order to compile a work of reference that leaves few, if any, questions unanswered.

1995   800 pages   0 566 07612 8

Gower

Gower Handbook of Marketing
Fourth Edition

Edited by Michael J Thomas

This new edition of a well-established Gower Handbook has been extensively revised and updated. Numerous chapters have been added, on subjects as diverse as relationship marketing and international marketing research, and there are many new contributors.

Part 1 reflects the need for a strategic view of the marketing function and looks in detail at information systems, planning, environment analysis and competitor analysis. Part 2 covers the organization of marketing, including recruitment, training, brand management and finance. Part 3 looks at product development (including services), and Part 4 with distribution. The final Part examines a number of aspects of marketing where new developments are making a profound impact and casts fresh light on such familiar topics as advertising, sales promotion, direct mail and franchising.

The 36 contributors represent an immense range of expertise. They are all acknowledged leaders in their chosen field, with practical experience of marketing.

1995   672 pages   0 566 07441 9

Gower

Licensing
The International Sale of Patents and Technical Knowhow

Michael Z Brooke and John M Skilbeck

This book is designed to take the reader through the maze of activities necessary for the successful selling of technical expertise internationally. It provides a comprehensive review of licensing for the practitioner: how and where licensing is used, the kinds of business supported, the opportunities, the problems and their solutions, together with other relevant issues.

After Part 1, which summarizes current usage, Part 2 examines the strategic aspects of licensing as a method of operating outside the home country; the relevant decisions are listed as are other options such as investment and franchising.

In Part 3 the authors turn to legal and political issues and include a specimen agreement. Part 4 deals with the managerial issues – including organizing, planning, financing, marketing and staffing – and concludes by examining the vexed question of relationships between licensor and licensee.

Part 5 looks at special considerations for particular nations and regions (including the developing world) while Part 6 summarizes and looks to the future.

The result is a comprehensive and up-to-date view of the issues and questions that face the licensing executive, together with practical guidance on dealing with these issues effectively.

1994   452 pages   0 566 07461 3

Gower

Professional Report Writing

Simon Mort

Professional Report Writing is probably the most thorough treatment of this subject available, covering every aspect of an area often taken for granted. The author provides not just helpful analysis but also practical guidance on such topics as:

- deciding the format
- structuring a report
- stylistic pitfalls and how to avoid them
- making the most of illustrations
- ensuring a consistent layout

The theme throughout is fitness for purpose, and the text is enriched by a wide variety of examples drawn from the worlds of business, industry and government. The annotated bibliography includes a review of the leading dictionaries and reference books. Simon Mort's book is destined to become an indispensable reference work for managers, civil servants, local government officers, consultants and professsionals of every kind.

Contents

Types and purposes of reports • Structure: introduction and body • Structure: conclusions and recommendations • Appendices and other attachments • Choosing words • Writing for non-technical readers • Style • Reviewing and editing • Summaries and concise writing • Visual illustrations • Preparing a report • Physical presentation • Appendix I Numbering systems • Appendix II Suggestions for further reading • Appendix III References • Index.

1992   232 pages   0 566 02712 7

Gower